EURI

And Other Stories.

A Book of Scientific Anecdotes

Also by Adrian Berry

Non-fiction

The Next Ten Thousand Years
The Iron Sun
From Apes to Astronauts
The Super-Intelligent Machine
High Skies and Yellow Rain
Ice with Your Evolution

Fiction

The Fourth Reich*
The Empire in Arumac
Koyama's Diamond
Labyrinth of Lies

Computer Software

Stars and Planets
The Wedgwood Benn Machine
Kings and Queens of England
Secret Key (*with Keith Malcolm*)

*Under the pseudonym Martin Hale

EUREKA!

And Other Stories:

A Book of Scientific Anecdotes

Chosen and introduced by

Adrian Berry

Helicon

First published by Harrap Books 1989
Second edition 1993

Helicon Publishing Ltd
42 Hythe Bridge Street
Oxford OX1 2EP

Printed and bound in Great Britain by
The Bath Press, Bath, Avon

ISBN 0 09 178276 7

British Library Cataloguing in Publication Data

A catalogue record for this book is available
from the British Library

To Laurence

THE GOOD effects wrought by founders of cities, law-givers, fathers of the people, extirpers of tyrants, and heroes of that class, extend but for short times; whereas the work of the Inventor, though a thing of less pomp and show, is felt everywhere and lasts for ever.

– Francis Bacon

Contents

Acknowledgements   10
Introduction   11

1 Tales of Explorers
Genius and Madness   15
Southward the Caravels   16
Magellan Enters the Pacific   20
The Amazon and its Amazons   22
Murder in the Bull-ring   28
'The Underground Rat of the North'   30
Scott's Last Journey   33
The First Free Diver   35

2 Air and Space
Too Close to the Sun   38
'The Dangerous Deserts of Space'   40
Faster than Sound   44
'Call Me Doctor!'   45
We Came in Peace for All Mankind   46
'We've had a Problem'   49
Challenger and Chernobyl   56

3 The Deep Skies
The World Moves!   64
The Story of Tycho Brahe   68
The Misdeeds of Isaac Newton   71
Nothing for Nothing   72
'The Learned Astronomer'   73
Einstein and Time Travel   74
'I'll Bring You to Your Senses!'   80
The Lonely Life of a Double Planet   82
The Cosmic Network   86

4 Communicating
Telegraph Hill   91
The Conquest of Solitude   94

Stop Screwing and Answer That Phone!   96
Glossary of Incompetence   96
Death by Secret Writing   98
The Unbreakable Cipher   101
The Badge of Cain   108
Get Rid of those Knobs and Levers   110
Can a Machine Think?   112
Sticking in the Knife   119

5 Ancestors
Darwin at the Galapagos   122
Finding Lucy   128
That Infuriating Smile   135
The Hard and Woolly Sciences   135

6 Martyrs
Murder of a Young Scientist   139
Victim of Revolution   140
Genius Against Stupidity   145
Medical Heroes   151

7 Bogus Science
The False Authority   156
'The Hollow Earth'   157
The Lunatic Fringe   158
A Bloody Bore   163
The Bogus Dentist   164

8 Utilities
Kaldi and the Coffee Bean   167
The Reckoning Master   169
The Light-Mile   173

9 War
A Delay in the Taking of Syracuse   175
'Those Wondrous New-made Bombards'   179
Einstein and the Bomb   181
'Expert' Opinion   188

10   Science Fiction
Why It Comes True   189
The New Atlantis   191

11   Diverse Matters
Bacon's Revolution   198
Napoleon and Science   199
A Panting Recitation   201
But Who Are You?   201
The Conquest of Smallpox   202
'Microbes are a Menace'   203
'Lousy' School Books   206
Sharks   211
The Baron and the Thief   214
Heaven is Hotter than Hell   217
The Explosions Within Us   218
Murphy's Laws of Technology   226

Appendix
The March of Knowledge   228

Bibliography   239
Quoted material: acknowledgements   244
Index   248

Acknowledgements

Many people helped me to write this book. I particularly thank Dr Anthony Michaelis for contributing so much of his knowledge, my wife Marina for reading through the manuscript and making many useful suggestions, and Matt Pritchett of *The Daily Telegraph* for his splendid drawings (including his cartoon on the cover).

Grateful thanks are due also to Nicholas Berry, Gulshun Chunara, David Atkins, Derek Johns, Susanne McDadd, Ian Hyde, Roy Minton, Roger King, Douglas Matthews, Gordon Hopkins, Enda Jackson, Linda Kelly, Laurence Kelly and Eleanor Berry.

Nor could I have made such progress without the helpful assistance of the staff of the Science Museum Library in South Kensington, of the London Library, and of the library of the Royal Astronomical Society.

Introduction

In the hundred thousand years since the appearance of the modern human brain, only the last four hundred have seen the truly modern *mind*. And only in the last fifty have technological advances accelerated to that headlong pace that seems destined to carry our descendants to the stars.

'This 'ere progress,' says a character in one of H.G. Wells's novels, 'it's wonderful 'ow it keeps going on.' We are fortunate that it does so; for were it ever halted or put into reverse, the consequences would be dire. Without the ever-accelerating acquisition of new knowledge and new techniques, humanity would eventually become extinct. The stage has passed when it was correct to say that we 'possessed' science. It is now nearer the truth to say that it possesses us.

Yet for the first four thousand years of recorded history, from the rise of the Sumerians to the end of the Dark Ages, there was little enthusiasm for improvement. The mind — or rather man's attitude to the Earth and to the Universe — seems to have gone through three stages, each distinct from the last.

The Dark Ages are generally reckoned to have begun at the fall of the Western Roman Empire, in 476 AD. But as far as science is concerned, one would put their start at the *beginning* of the Empire. After the Battle of Actium in 31 BC, when the future

Emperor Augustus established absolute dominion over the known world, intellectual and material progress came to a halt that lasted some thirteen centuries. As Blaise Pascal said of that fatal little war between Rome and Egypt, 'Had Cleopatra's nose been shorter, the whole history of the world might have been different.'

The Romans made virtually no progress that the Greeks had not made before them. They created an imposing if oppressive architecture of the kind that Hitler long afterwards wished Albert Speer to recreate in Germany. Their central heating, their luxurious baths and even their shorthand were all Greek — or sometimes Cretan — inventions. It is true that without the guidance of any earlier civilization they learned to build aqueducts and military roads, and they were admittedly experts in the concoction of poisons. But these were the limits of their technical achievements. Relying on slaves, they had no incentive to build machines, and their cavalry even rode without stirrups. They had little curiosity. They revered authority rather than knowledge. Who cared about the size and distance of the Moon so long as one could steer and march by its light? They shrank from the unknown at every encounter. Mariners feared to sail out of sight of land; 'the hungry sea is fatal to sailors,' wrote Horace shudderingly. The planets were deified, comets regarded as portents and the stars only examined to reveal fortunes. Such science that existed was equated with philosophy, a vocation incapable of usefulness and hence despised.

It was a world that could have had no future. 'I opposed the dictatorship of Sulla,' said Julius Caesar, 'because his was a dead world.' He might have said the same about the centuries that followed. Even the most successful period of the Roman Empire, the reigns of the five Antonine Emperors in the second century AD, was a period of almost absolute mental stagnation. People seemed to take a complacent satisfaction in knowing next to nothing. 'I will ignore,' said Julius Frontinus, Trajan's leading military engineer, 'all new ideas for works and engines of war, whose invention has reached its limits and for whose improvement I see no hope.'

A veneer of prosperity concealed the consequences of this worship of ignorance. 'It was scarcely possible,' wrote Edward Gibbon,

'that the eye of contemporaries should discover in the public felicity the latent causes of decay and corruption. The long

peace, and the uniform government of the Romans, introduced a slow and secret poison into the vitals of the empire.
The minds of men were gradually reduced to the same level,
the fire of genius was extinguished, and even the military spirit
evaporated. The diminutive stature of mankind was daily
sinking and the Roman world was peopled by a race of
pygmies, when the fierce giants of the north broke in and
mended the puny breed. They restored a manly spirit of
freedom, and, after a revolution of ten centuries, freedom
became the happy parent of taste and science.'

This Renaissance, as the science-fiction writer James P. Hogan has
argued, was misnamed. It was not the *re-birth* of anything, but
rather as if a second species had come into existence. The whole
direction of thought was the opposite to what it had been. The great
discovery of Copernicus, the cosmology of Newton and the mathematics of Leibnitz 'was the language of a race no longer static and
bounded but one dynamic and unlimited, to be explored through a
passion for discovery'.
  In the voyages of the early explorers alone, the pace of advance
became literally frenzied. In the sixteenth century, wrote Stefan
Zweig,

> The picture of the world altered from year to year, nay from
> month to month. By day and by night the cartographers were
> hard at work in Augsburg, but could not keep pace with the
> unceasing demand for revised maps. They were snatched from
> their hands damp and still uncoloured. Hardly had they
> revised them in accordance with the latest intelligence when
> new reports arrived. The old representations were thrown
> aside, and fresh ones made, for what had been an island was
> now known to be a continent; a continent in which new rivers,
> new mountains had to be depicted. And the etchers had
> scarcely refurbished their maps when yet more news came to
> hand demanding still further revision.

What a contrast to the situation reported by Gibbon! The social tide
was now running in the opposite direction. Science soon came to
embrace not only exploration but nearly every facet of existence.
'Its once feeble vanguards,' wrote Winston Churchill in 1932,

'often trampled down, often perishing in isolation, have now become a vast organized army marching forward on all fronts towards objectives that none may measure or define. It is an army that cares nothing for all the laws that men have made; nothing for their most time-honoured customs, nothing for their most dearly-cherished beliefs, nothing for their deepest instincts.'

This very *unexpectedness* of science is now generating a new crisis. A new mental species is appearing. Science has become so vast, complicated and frightening that new attitudes are needed to comprehend even small parts of it. The developments of this century — relativity, quantum mechanics, nuclear power, genetic engineering, space travel, electronics and scientific agriculture and the beginnings of artificial intelligence — have swept away most of the predictions of the last. Who, for example, would have dared to predict only fifty years ago that hundreds of filing clerks could be replaced by a single personal computer, able to store the equivalent in data of forty novels? Who would have suggested that steamships would be superseded almost entirely by jet aircraft that travelled thirty times faster? And what, above all, do the example of these advances augur for the twenty-first century?

This book attempts to illuminate some of those inventions and discoveries that have changed the world, and the people who made them.

It is a highly personal collection, and in no sense intended to be a 'complete' record of all the important events in the history of science. Were it so it might be so bulky that it would need one of Archimedes's pulleys to lift it. Why have I described *this* and not *that*? The answer, in most cases, is that I could find no interesting eye-witness or later historical account. There is nothing here, for example, about the Wright brothers' first manned flight in 1903. Why? Because the local newspaper editor — the kind of person who finds science incredible — would not send a reporter to cover 'this ridiculous story'.

ADRIAN BERRY
London, 1989

1
Tales of Explorers

Genius and Madness

Christopher Columbus, like many great explorers, came to an evil end, as Timothy Ferris relates:

Towards the end of his life Columbus roamed the coasts of the New World in a state of gathering madness. He kept a gibbet mounted on the taffrail of his ship from which to hang mutineers, and made use of it so frequently that at one point he had to be recalled to Cadiz in chains. Crewmen on his final voyage watched warily as their captain hobbled around the deck, his body twisted by arthritis, his wild eyes peering out from under an aurora

of tangled hair, searching endless coastlines for the mouth of the River Ganges. He threatened to hang anyone who denied they were in India. He sent back shiploads of slaves, which alarmed his Queen,* and cargos of gold, which delighted them both. "O, most excellent gold," Columbus wrote. "Who has gold has a treasure with which he gets what he wants, imposes his will in the world, and even helps souls to paradise." He died poor.

Southward the Caravels

What is Alexander crowned with trophies at the head of his army, compared with Henry contemplating the ocean from his window on the rock of Sagres?

— W. J. Mickle, Preface to his translation of Camoen's epic poem *The Portuguese.*

Europe's colonization of the world that began in the fifteenth century was started by a single determined genius, Prince Henry the Navigator of Portugal. This man, who never set foot in a ship (apart from a short military expedition), left behind no book or journal. But he used his authority as a prince of the blood, and his income as the Grand Master of the Order of Christ, to direct the exploration of the unknown Ocean of Darkness — the Atlantic — and the western coast of Africa.

From his headquarters at Sagres, a lonely port on the southern tip of Portugal, he sent forth countless expeditions. In these he shattered the myths that had been propagated of old by the astronomer Ptolemy, that a ship that sailed too far to the west would fall off the edge of the world, and that one that ventured too far south would be roasted by the vertical rays of the sun.

Years later the Italian poet Politian made this acknowledgement of the debt which Europe owed to Portugal:

Not only has this country left far behind the Pillars of Hercules and curbed a raging ocean, but it has reestablished the hitherto interrupted unity of

* Isabella of Castile.

16

the habitable globe. What fresh possibilities and economic advances, what an enlargement of knowledge, what confirmation of old science hitherto rejected as incredible, may now be expected! New countries, new seas, new worlds, have emerged from the darkness of the centuries. Portugal has become the guardian of a second universe.

But with what kind of ships were these feats achieved? Henry's biographer Ernle Bradford gives us this account:

The ships of the earliest expeditions were barks or *barinales*. They were square-sailed vessels of the type that normally carried river cargo. They were admirable for running *before* the wind, but Prince Henry's mariners had good reason to fear that they could not beat back *against* it. The difficulty that they experienced in returning to Portugal against the prevailing north winds led to the evolution of the famous caravel.

Like most ships, the caravel had a long ancestry. Even today on the Tagus, one can see small sailing boats that have been claimed as its descendants. These are broad-beamed, shallow-draught craft, pine-planked on oak frames. Sometimes, at sea off the Portuguese coast, one sees them stooping and soaring over the Atlantic rollers, their triangular lateen sails curved above a short mast. They are open, or half-decked boats. It is very likely that the caravels of Prince Henry were derived from the ancient Arab dhows of the Red Sea.

An early 16th century painting shows a vessel that is probably similar to the caravels. The stem is quite graceful, almost yacht-like, with a certain amount of overhang. The general line of the bulwarks shows a gentle sheer. The stern is built up into a poop with an aftercastle, in which is stepped a mizzenmast. The mainmast is a little behind the centre line of the vessel, and there are two lateen sails. The sails are easily controlled, and one or two men can handle quite a large area of canvas.

Caravels had no topsails to worry about. They had no complicated system of braces as did square-rigged ships, and no intricate sheeting of sails. Until the introduction of the fore-and-aft sail many centuries later, the lateen was the most efficient rig for small boats. As the Venetian Cadamosto remarked of Prince Henry's caravels, they were "the most seaworthy vessels of their time."

They usually weighed less than 100 tons. Historians often seem to have been amazed at the smallness of these vessels in which the early navigators made their ocean voyages. In fact they were well designed and seaworthy, with a good, easily worked rig. Size has little to do with seaworthiness. (Yachts of ten tons and less quite often cross the Atlantic nowadays.)

The pine of which they were built had long been a protected tree in Portugal. This was partly because that country, without its defensive barrier of pine trees, would otherwise have been invaded by sand, driven inshore on the Atlantic winds. These native stone pines were particularly suitable for ribs, strakes, and curved pieces. They also provided good straight wood for planking. As in England, oak was used for the keels. Lisbon, Porto and Sagres were the main shipbuilding centres, noisy with the thud of adzes on wood, and lively with attendant trades like sailmaking and ropemaking.

Even before the death of Prince Henry in 1460, the caravels had increased in size, and when they exceeded 100 tons, they often had a third and even a fourth mast. These later vessels had a composite rig — lateen on the mizzen and main masts, and square sails on the foremast. Sixty to 90 feet long, with a beam of 20 to 30 feet, they were well-proportioned ships. Their broad beam allowed them to be comparatively shallow-drafted, and this was to prove an enormous advantage when working along the African shore in uncharted and often shallow waters.

In sailing close to the wind, the caravel was far superior to any vessel of its time. This enabled Prince Henry's ships not only to coast down Africa, but to sail back again against the prevailing winds. The sailors no longer felt that they were condemned to run before the wind without hope of return.

What instruments and charts had these early seafarers? For centuries, Mediterranean sailors had navigated in coastal waters in sight of land. They had reasonably accurate charts, and they relied on those three stand-bys of pilotage — log, lead, and look-out. For centuries too, seamen had been running between the Mediterranean and the northern countries of Europe. But when it came to venturing south into the Atlantic, out of sight of land, along a coastline of which there was no verbal record — let alone any charts — accurate knowledge of latitude became all-important. They had to plot their course by compass.

The simple magnetic compass in a caravel was kept in a small wooden binnacle, lit at night by a lamp burning whale or olive oil. Often enclosed in the binnacle itself, was the hourglass — the only time-keeping device then available at sea. Clocks were still rarities, and totally unsuited for use in a small vessel. A ship's boy kept watch alongside the helmsman. It was his duty to maintain the lamps and the binnacle light, as well as to watch the hourglass; reversing it when the sand had run out and calling out the hour. A grave offence, and punishable by flogging, was the trick of "warming the glass," when the boy either held the hourglass near the lamp or put it inside his shirt to warm. The result was that the thin glass expanded, the sand ran through faster — and the long, tedious watch was shortened.

They were familiar also with the constancy of the Northern Star as a reference point, used for navigation since classical times. It was known as the Star of the Sea, a name often given to the compass itself. Felix Faber, a monk writing about 20 years after Prince Henry's death, described the steering in a voyage he made to Palestine:

"They have a Star of the Sea near the mast, and a second one on the topmast deck of the poop. Beside it all night long, a lantern burns. A man constantly watches the Star, never taking his eyes off it. He sings out a pleasant tune, telling that all goes well, and in the same chant directs the man at the tiller. The helmsman dares not move the tiller in the slightest degree, except at the orders of the man who watches the Star of the Sea."

Reasonably accurate charts of the Mediterranean and of Europe's Atlantic seaboard had been current for many years. The oldest surviving maritime chart is the Carta Pisana dated about 1275. By the early 14th century the chartmakers of Genoa were well known throughout Europe, and the Catalan Jews of Majorca were equally famous for their knowledge and draftsmanship. An outstanding figure of the period just before Prince Henry was Abraham Cresques of Majorca, described as a "master of charts and compasses." His son Jaime was one of the leading cartographers whom Henry induced to settle at Sagres. Jaime Cresques joined the

Prince's court a year or so after the discovery of Madeira. From then on he was engaged in research, in chart making, and in helping the Prince to correlate the information brought in by his sea captains. There is no doubt that he would also have brought with him from Majorca mathematical tables, similar to those mentioned by Ram Lull a century earlier, with which — by the calculation of rhumb lines — sailors could work out their mileage while at sea.

Henry's captains and navigators had in the seaman's astrolabe a fairly efficient instrument for calculating their positions by the altitudes of the sun and the stars. Like so much else connected with the mathematical sciences, the astrolabe may have originated with the Arabs. We know from Marco Polo in the early 14th century that the Arabs had long been using a simple instrument called the *kamal* for star observations.

The astronomer's astrolabe was a costly and elaborate but accurate instrument. It was well known by the 15th century, but was unsuited for use at sea. From it was developed a seaman's version, with a simple scale of degrees for measuring the height of the sun or a star. It was this type of instrument that the Portuguese used in their Atlantic and African explorations.

This was the beginning of a new era of navigation. At Lisbon or Sagres, the navigator became familiar with his tools, and measured the altitude of the sun or stars at his home port. Then, as he sailed south, as he discovered new capes, headlands, and bays, he worked out his difference in latitude.

The Portuguese navigators still hugged the coast as much as possible. Their instruments were not reliable, they were constructing their charts as they went along, and their shallow-draft vessels allowed them to work close inshore. When they came to a new point or headland, they would go ashore with astrolabe and tables, and take the sun's altitude at noon from dry land. The heaving deck of a caravel was no place for accurate measurement with an awkward instrument.

These were the ships and the navigational aids with which the Portuguese opened up the world. With each new voyage, their captains gradually learned their deep-sea trade, and their crews grew used to the wide spaces of the unknown ocean.

Magellan Enters the Pacific

Ferdinand Magellan of Portugal, the first explorer to sail almost round the world, faced difficulties that would have thwarted any but the most determined. Christopher Columbus had been unable to reach the Indies by sailing west, and Magellan intended to succeed where he had failed. But like Columbus before him, he took long years to get the backing that he needed for such a voyage.

Spurned by his king, Manuel I (who apparently took a personal dislike to him), Magellan offered his services to Spain. Spain was in a strong position to take advantage of them, for shortly after the first voyage of Columbus in 1492, Pope Alexander VI had drawn a north–south line down the Atlantic. All the lands discovered to the west of this line were to belong to Portugal, and all east of it to Spain. Magellan therefore pointed out to the Spanish king that if he could reach the Indies he could lawfully seize them for the Spanish crown.

Even this argument nearly failed, for the Spanish court could think only of European politics. At last Magellan got his backing with the unexpected aid of Fonseca, the Bishop of Burgos, who had already made himself ridiculous by opposing the projects of Columbus, and had no wish to do so a second time.

Magellan retained only one secret: a chart he had obtained in Portugal, which had been drawn up by Peter Behaim, or Martin the Bohemian. It showed a navigable route westward through the tip of South America to the still unknown Pacific. This route was later to be called the Strait of Magellan.

At length, in September 1519, he set out across the Atlantic with five ships. He had been harassed by Portuguese agents who accused him of disloyalty to his country, and he was to face a mutiny by two of his captains which he ruthlessly suppressed. But Magellan had with him that most precious asset of an explorer, a diarist and chronicler, Antonio Pigafetta, whose journals are the main source for his great voyage.*

In the following passage, Pigafetta describes Magellan's crowning triumph, the actual discovery of the hidden strait. It was not in the place shown in Behaim's map, and nearly all in the expedition by Magellan had despaired of finding it. But the morning of 21 October 1520 marked a mighty episode in the history of exploration:

After setting course to the fifty-second degree towards the Antarctic Pole, on the festival of the eleven thousand virgins, we found by miracle a strait

---

* Pigafetta's description of some of the natives of Patagonia inspired Shakespeare's creation of Caliban in *The Tempest*.

which we called the Cape of the Eleven Thousand Virgins. This strait is four hundred and forty miles long, and somewhat less than two miles wide. It is surrounded by very great and high mountains covered with snow.

It was impossible to anchor, because no bottom was found. We therefore secured the ships by cables to the shore. There seemed to be no way out of this strait. But the Captain-General [Magellan] insisted that there was another strait that led out of it to the Pacific Sea, saying that he had seen it in a chart in the treasury of the King of Portugal, drawn by a great pilot and sailor named Martin Behaim.

The Captain-General sent forward two of his ships, one named *San Antonio* and the other *Concepcion*, to seek the outlet of this strait. And we with the two other ships, the flagship *Trinidad* and the *Victoria*, remained waiting. And in the night we had a great storm which lasted until noon of the next day. We were compelled to loose the cables and let our ships toss hither and thither.

The other two ships, meanwhile, had such a stormy passage that they could not go forward. Trying to return to us, they were hard put not to run aground. Thinking themselves lost, they found a small opening which seemed but a creek. Like desperate men they threw themselves into it, so that perforce they discovered the strait beyond.

Seeing that this was not a creek, but the opening to another strait, they went on through it, and found a bay that opened into the Pacific Sea. Very joyful, they at once turned back to inform the Captain-General.

We thought they had perished, because of the great storm and because we had not seen them for two days. We had also seen smoke signals [which might have meant shipwreck and a call for help.] And while in suspense, we saw the two ships approaching us under full sail and flying their banners. When near us, they suddenly fired their guns. We very joyously greeted them in the same way. And then we all went forward together, thanking God and the Virgin Mary.

— Antonio Pigafetta

The Amazon and its Amazons

Few tales in history are darker and more splendid than the story of how a handful of Spanish adventurers, avid for gold, seized and destroyed the empire of the Incas. Not content with plundering

Peru of its gold and silver, they lusted after a legendary kingdom of El Dorado, a land of gold, which was supposed to exist somewhere in the forests of the Amazon.

William Prescott, one of the most inspiring of nineteenth-century historians, describes here how Gonzalo Pizarro, younger brother of the Conqueror of the Incas, set out after the Conquest to find El Dorado:

Gonzalo Pizarro received the news of his appointment to the government of Quito [in 1539] with undisguised pleasure: not so much for the possession it gave him of this ancient Inca province, as for the field it opened for discovery towards the east, the fabled land of Oriental spices, which had long captivated the imagination of the Conquerors. In a short time, he mustered 350 Spaniards and 4,000 Indians. One hundred and fifty of his company were mounted, and all were equipped in the most thorough manner for the undertaking. He provided, moreover, against famine by a large stock of provisions and an immense drove of swine which followed in the rear.

It was the beginning of 1540 when he set out on this celebrated expedition. The first part of the journey was attended by comparatively little difficulty, while the Spaniards were yet in the lands of the Incas: for the distractions of the Conquest in Peru had not been felt in this distant province, where the simple people still lived as under the primitive sway of the Children of the Sun.

But the scene changed as they entered the territory of Quixos, where the character of the inhabitants, as well as of the climate, seemed to be of another description. The country was traversed by the lofty ranges of the Andes, and the adventurers were soon entangled in their deep and intricate passes. As they rose into the more elevated regions, the icy winds that swept down the sides of the Cordilleras benumbed their limbs, and many of the natives found a wintry grave in the wilderness. While crossing this formidable barrier, they experienced one of those tremendous earthquakes which, in these volcanic regions, so often shake the mountains to their base. In one place, the earth was rent asunder by the terrible throes of Nature, while streams of sulphurous vapour issued from the cavity, and a village with some hundreds of houses was precipitated into the frightful abyss!

On descending the eastern slopes, the climate changed: and, as they came on the lower level, the fierce cold was succeeded by a suffocating heat, while tempests of thunder and lightning, rushing out from the gorges of the sierra, poured on their heads with scarcely any intermission day or night, as if the offended deities of the place were willing to take vengeance on the invaders of their mountain solitudes. For more than six weeks the deluge continued unabated, and the forlorn wanderers, wet and weary with incessant toil, were scarcely able to drag their limbs along the soil broken up and saturated with the moisture. After some months of toilsome travel, in which they had to cross many a morass and mountain stream, they at length reached *Canelas*, the Land of Cinnamon. They saw the trees bearing the precious bark spreading out into broad forests: yet, however valuable an article for commerce it might have proved in accessible situations, in these remote regions it was of little worth to them. But, from the wandering tribes of savages whom they occasionally met in their path, they learned that at ten days' distance was a rich and fruitful land abounding with gold, and inhabited by populous nations. Gonzalo Pizarro had already reached the limits originally proposed for the expedition. But this intelligence renewed his hopes, and he resolved to push the adventure farther. It would have been well for him and his followers had they been content to return on their footsteps.

Continuing their march, the country now spread out into broad savannas terminated by forests which, as they drew near, seemed to stretch away on every side to the very verge of the horizon. Here they beheld trees of that stupendous growth seen only in the equinoctial regions. Some were so large that sixteen men could hardly encompass them with extended arms!* The wood was thickly matted with creepers and parasitical vines, which hung in gaudy-coloured festoons from tree to tree, clothing them in a drapery beautiful to the eye, but forming an impenetrable network. At every step of their way, they were obliged to hew open a passage with their axes, while their garments, rotting from the effects of the drenching rains to which they had been exposed, caught in every bush and bramble, and hung around them in shreds. Their provisions, spoiled by the weather, had long since failed, and the livestock which they had taken with them had either been consumed or made their escape in the woods and mountain passes. They had brought with them nearly a thousand dogs, many of them of the ferocious breed used in hunting down the unfortunate natives.** These they now gladly killed, but their miserable carcasses furnished a lean banquet for the famished travellers: and, when these were gone, they had only such herbs and dangerous roots as they could gather in the forest.

At length the way-worn company came on a broad expanse of water formed by the Napo, one of the great tributaries of the Amazon, and which, though only a third or fourth rate river in America, would pass for one of the first magnitude in the Old World. The sight gladdened their hearts, as,

---

* Allowing six feet for the spread of a man's arms, this would make these trees about 96 feet in circumference. They fall short only of that stupendous tree in the Oaxaca state of Mexico which is a full 112 feet in girth. — A.B.

** The Conquistadors were also in the habit of using these dogs to tear natives to pieces as punishments. — A.B.

by winding along its banks, they hoped to find a safer and more practical route. After traversing its borders for a considerable distance, closely beset with thickets which it taxed their strength to the utmost to overcome, Gonzalo and his party came within hearing of a rushing noise that sounded like subterranean thunder. The river, lashed into fury, tumbled along over rapids with a frightful velocity, and conducted them to the brink of a magnificent cataract which, to their wondering fancies, rushed down in one vast volume of foam to the depth of twelve hundred feet! The appalling sounds which they had heard from the distance of eighteen miles were rendered yet more oppressive to the spirits by the gloomy stillness of the surrounding forests. The rude warriors were filled with sentiments of awe. Not a bark dimpled the waters. No living thing was to be seen but the wild tenants of the wilderness, the unwieldy boa, and the loathsome alligator basking on the borders of the stream. The trees towering in wide-spread magnificence towards the heavens, the river rolling on in its rocky bed as it had rolled for ages, the solitude and silence of the scene, broken only by the hoarse fall of the waters, or the faint rustling of the woods: all seemed to spread out around them in the same wild and primitive state as when they had come from the hands of the Creator.

For some distance above and below the falls, the bed of the river contracted so that its width did not exceed twenty feet. Sorely pressed by hunger, the adventurers determined, at all hazards, to cross to the opposite side, in the hopes of finding a country that might afford them sustenance. A frail bridge was constructed by throwing the huge trunks of trees across the chasm, where the cliffs, as if split asunder by some convulsion of nature, descended sheer down a perpendicular depth of several hundred feet. Over this airy causeway, the men and horses succeeded in affecting their passage with the loss of a single Spaniard who, made giddy by heedlessly looking down, lost his footing and fell into the boiling surges below.

Yet they gained little by the exchange. The country wore the same unpromising aspect, and the riverbanks were studded with gigantic trees or fringed with impenetrable thickets. The tribes of Indians, whom they occasionally met in the pathless wilderness, were fierce and unfriendly, and they were engaged in perpetual skirmishes with them. From these they learned that a fruitful country was to be found down the river at the distance of only a few days' journey, and the Spaniards held on their weary way, still hoping and still deceived, as the promised land flitted before them like the rainbow, receding as they advanced.

At length, spent with toil and suffering. Gonzalo resolved to construct a bark large enough to transport the weaker part of his company and his baggage. The forests furnished him with timber: the shoes of the dead horses were converted into nails: and the tattered garments of the soldiers

25

supplied a substitute for oakum. It was a work of difficulty: but Gonzalo cheered his men in the task, and set an example by taking part in their labours. At the end of two months a brigantine was constructed, rudely put together, but strong and of sufficient burden to carry half the company: the first European vessel that ever floated on these inland waters.

Gonzalo gave the command of the bark to Francisco de Orellana, a cavalier on whose courage and devotion to himself he thought he could rely. The troops now moved forward, still following the descending course of the river, while the brigantine kept alongside: and when a bold promontory or more impracticable country intervened, it furnished timely aid by the transportation of the feebler soldiers. In this way they journeyed, for many a wearisome week, through the dreary wilderness of the borders of the Napo. Every scrap of provisions had long since been consumed. The last of the horses had been devoured. To appease the gnawings of hunger, they were fain to eat the leather of their saddles and belts. The woods supplied them with scanty sustenance, and they greedily fell upon toads, serpents, and such other reptiles as they occasionally found.

Here we must leave Gonzalo Pizarro and follow the adventures of Orellana. That cavalier was sent off by Pizarro to sail down the river in the brigantine with a few soldiers to seek the provisions which the expedition so desperately needed. But he never returned. Some said that he found it too difficult to sail back again upstream: others that he treacherously abandoned his comrades and sailed off in search of the mythical land of gold.

What he actually did was even more fantastic. He joined the Amazon, sailed down it all the way to the Atlantic and crossed the ocean to Spain — a journey of more than six thousand miles in a ship that had been put together with old horse-shoes. The Spanish court listened with wonder to his tale of exploits, in particular to his battle with the 'Amazons,' a tribe of warrior women (named after a similar tribe in Greek mythology) who lived somewhere along its banks, and after whom the great river was later named. Robin Furneaux describes this mysterious encounter:

Rounding a bend in the river, they saw "many villages and very large ones, which shone white." Indians came out to meet them. Orellana talked to them in a peaceful manner, but they jeered and said that others were waiting for them downstream and would bind them and take them to the Amazons. As the Spaniards were again short of food, it seemed sensible to collect some at once, if they were to be faced with a long running fight on the river. But this raid nearly ended in disaster.

The village was defended by a horde of warriors, who resisted with ferocity. The bitter fight continued for an hour. The Spaniards noticed women fighting among the Indian ranks and using their war-clubs to strike down any warriors who fled. A Spaniard named Carvajal described them as "very white and tall with hair long and braided and wound about the head, and they are very robust and go about naked, but with their privy parts covered."

Throughout this province they were harried by the Indians, and given no chance to find food and little to rest. When they had left, Orellana at last found time to question an Indian trumpeter whom they had captured, and from him they learned the strange story of the Amazons.

He told them that the women who had fought against them lived in the interior of the country, and that many provinces were subject to them. Their villages were built of stone and inhabited only by women. When they wished, they would make war on a neighbouring overlord and capture husbands for themselves. If their children were female, they would raise them and train them in the arts of war, but if they were male they would be killed or returned to their fathers. They had much gold and silver, both as idols in their temples and as plates in their homes. Their appearance was as exotic as their possessions, magnificent blankets draped around them, or sometimes flung over a shoulder, golden crowns on their heads, and hair so long that it reached the ground. Animals, which the Spaniards guessed to be camels, carried them from place to place, and there were other unidentifiable beasts "as big as horses, and which had hair as long as the spread of the thumb and forefinger, and cloven hooves."

Apart from Carvajal's account of the Indian women leading their men into battle, this man's tale is the main authority for the legend of the Amazons. For centuries afterwards, many people believed it implicitly while others derided Orellana for giving credence to such a fantastic story.

At first sight the legend is founded entirely on the hearsay evidence of one frightened Indian. But the same story — of "the women who live alone" meeting men solely in order to breed — is part of Indian folklore throughout the Amazon and Orinoco basins and in the northern head-waters of the Paraguay. The stories of them told to Orellana and later to Sir Walter Raleigh were repeated with few changes to other explorers hundreds of years later. Mission Indians, in their confessions, claimed to have visited the Amazons and to have been rewarded by gifts of gold and green stones. One explorer, Richard Spruce, suggested that they may have been descendants of a fugitive vestal community from the empire of the Incas.*

Murder
in the Bull-ring

The great French explorer Charles-Marie de la Condamine is best known for his discovery in the Amazon Jungle of the tree sap called *caoutchouc*, which formed the basis of rubber. But before reaching the Amazon he was the leader, in 1735, of a disastrous expedition to Ecuador and Peru which was meant to disprove one of the theories of Isaac Newton. Robin Furneaux tells the story of its collapse:

Isaac Newton had advanced the theory that the Earth bulged at the equator and flattened at the poles. A group of French scientists, who thought it was elongated at the poles and attenuated in the middle, were irritated by foreign interference in a branch of science which they had always considered their own. The whole scientific world took sides in a dispute that became increasingly acrimonious until meetings of learned societies would dissolve in uproar. The French Academy of Sciences decided that the only way to settle the argument was for two expeditions to make the appropriate observations, one in the polar regions and one in the equator.

Charles-Marie de la Condamine, chronicler of the expedition to the equator and its most adventurous member, had been a soldier before becoming a scholar: he specialized in mathematics and geodesy, and could also act as a cartographer, astronomer or naturalist. His portrait shows us a good-looking face dominated by a lofty brow, an aquiline nose and large hooded eyes. Accompanying him were nine fellow-Frenchmen including a doctor, an astronomer, two mathematicians, a naval captain, a draughts-man, a watchmaker, and a boy, qualified to join such distinguished company by being the nephew of the Treasurer of the Academy.

The work of the Academicians was plagued by frustrations and tragedy. At Quito, they were fêted as heroes, but after a few days the novelty wore

---

* He also advances the attractive theory that the legend of El Dorado may have originated in stories of the wealth of Peru brought across the South American continent and back again in such a distorted form that the Spaniards failed to recognize them. If he was right, those who perished in the fruitless expeditions to the Amazon were searching for a kingdom which they had already conquered and which they were indeed inhabiting at the time. — R.F.

off and the Creoles set out to discover what the Frenchmen really wanted. There was only one South American answer to such a question, and that was hidden gold. All the meticulous measurements and incomprehensible scientific equipment helped to foster this suspicion, which became a certainty when the Academicians tried to explain what they were doing. Surely no one could be so stupid to cross half the world and spend miserable months in the icy highlands of the Andes just to look at the stars and measure the Earth's shape? Suspicious officials pestered them until work became impossible.

The expedition soon had more serious troubles. Couplet, the Treasurer's nephew, died of fever. Later, just as their work on the triangulations was being completed, they received the devastating news that the polar expedition had returned to France, having proved Newton's theory beyond any doubt. La Condamine found it hard to persuade his companions to continue.

They retired to Cuenca to rest, only to suffer the worst disaster of all. Dr. Senièrgues, a member of the expedition, became involved in a duel, having gallantly supported a girl who had been rejected by one of the town's most prominent citizens.

He met his enemy on a street corner. After words had been exchanged, Senièrgues drew his sword and the other man his pistol. The Frenchman charged. Too impetuous, he slipped on the cobbles. Before the affair could go any further, onlookers separated them.

A few days later Senièrgues appeared in a box in the bull-ring with the girl who was the cause of the feud. The whole town of Cuenca was now openly hostile. The master of the ring, another enemy, rode up and made some offensive remarks to which Senièrgues replied with such vigour that the man galloped away in panic to the jeers of the crowd. He then retaliated by announcing that, since his life had been threatened, the bullfight was cancelled.

At this the crowd rioted and headed for Senièrgues's box. He sprang out, his back to the barrier. His sword was in one hand, his pistol in the other. The mob, several hundred strong, pressed around him but were too cowardly to attack. Senièrgues started to edge along the barrier, darting his rapier at anyone who seemed too close. He had just reached the door when the mob hurled showers of stones. The mob engulfed him, stabbing, hacking and stamping until he was mortally wounded. La Condamine, other Frenchmen and a few sane Creoles managed to separate him from his murderers, place him on a litter and carry him out of the ring. The frustrated mob now stormed the box occupied by the remaining Frenchmen, forcing them to escape by a ladder.

Bouguer, the astronomer, and La Condamine were chased through the streets by the rabble, pelted with stones and forced to take refuge in a church. To complete the expedition's misery, de Jussieu, the botanist, lost all of his collection through the carelessness of a servant. The thought of the five years' privation that he had willingly undergone being wholly wasted was too much for him. Poor de Jussieu had a breakdown and never fully recovered his mental powers.

The expedition gradually disintegrated. La Condamine and Bouguer were no longer on speaking terms; and the former was being sued by the Spanish members of the expedition for omitting to put their names on the pyramids which he had erected to commemorate their work.

The Earth had been proved to bulge round the equator, but the pyramids controversy had made La Condamine universally loathed, and there was nothing now to keep him in the country.

'The Underground Rat of the North'

Huge animals, now extinct, lived during the last Ice Age, which ended about 11,000 years ago. Many people thought them mythical. But this view changed in the mid-nineteenth century, as Windsor Chorlton relates:

The Siberian spring of 1846 arrived with freakish and devastating swiftness. In May of that year, a Russian government survey team steamed from the East Siberian Sea into the mouth of the Indigirka River. Among its members was a young man known to history only as Benkendorf. The surveyors found the river in violent flood, swollen by melting snow and torrential rains that tore away its banks and carried great chunks of ice and frozen soil far out to sea. Abnormally high temperatures had sent melted water cascading across the entire north coastal region and had softened the surface of thousands of square miles of permafrost, turning the tundra into a treacherous morass which made an overland passage impossible.

Survey work could clearly not be carried out in these conditions, but the team nevertheless decided to venture upstream in their small steamboat. The only existing story of the journey, and the extraordinary discovery to which it led, was recorded by Benkendorf:

"Steaming up the Indigirka, we saw no signs of land. The landscape was flooded as far as the eye could see. We saw around us only a sea of dirty brown water, and knew that we were in the river only by the strength of the current. A lot of debris was coming downstream, uprooted trees, swamp litter and large masses of tufts of grass, so that navigation was not easy."

After toiling against the current for eight days, the surveyors reached the spot where they had arranged to met the local guides who, not surprisingly, failed to appear.

"We knew the place. But how it had changed! The Indigirka, here about two miles wide, had torn up the land and made itself a fresh channel. When the floods subsided we saw to our astonishment that the old riverbed had become merely that of an insignificant stream. We explored the new stream, which had cut its way westward. We later landed on the new bank, and surveyed the undermining and destructive work of the wild waters that were carrying away, with extraordinary rapidity, masses of peat and loam.

"The stream was tearing away the soft sodden bank like chaff, so that it was dangerous to go near it. We suddenly heard a gurgling and movement in the water under the bank. One of our men gave a shout, and pointed to a shapeless mass which was rising and falling in the swirling stream. We all hastened to the bank. We had the boat brought up close, and waited until the mysterious thing should again show itself.

"At last a huge black horrible mass bobbed up out of the water. We beheld a colossal elephant's head, armed with mighty tusks, its long trunk waving uncannily in the water, as though seeking something it had lost. Breathless with astonishment, I beheld the monster with the whites of its half-open eyes showing. 'A mammoth! A mammoth!' someone shouted."

Fetching chains and ropes, the excited surveyors attempted to secure the carcass of the mammoth before the river carried it away, and after many tries they managed to get a line around its neck. Benkendorf then realized that the hindquarters of the beast were still embedded in the frozen riverbank, and he decided to let the river finish excavating the carcass before attempting to haul it on to dry land. A day passed before the beast was free of the ice, and during the wait the local guides arrived on horseback. With the aid of the horses and the extra men, the crew dragged the mammoth ashore and pulled it away from the riverbank. For the first time, Benkendorf was able to get a good look at the animal:

"Picture to yourself an elephant with a body covered with thick fur, about 13 feet in height and 15 feet in length, with tusks eight long, thick and curving outwards at their ends. A stout trunk six feet long, colossal legs one and a half feet thick, and a tail bare up to the tip, which was covered with thick, tufty hair. The beast was fat and well-grown. Death had overwhelmed him in the fullness of his powers. His large, parchment-like, naked ears lay turned up over the head.

"About the shoulders and back he had stiff hair about a foot long, like a mane. The long outer hair was deep brown and coarsely rooted. The top of the head looked so wild and so steeped in mud that it resembled the ragged bark of an old oak tree. On the sides it was cleaner, and under the outer hair there appeared everywhere a wool, very soft, warm and thick, of a fallow brown tint. The giant was well protected against the cold."

The mammoth began to decay as soon as it came free from its frozen tomb. Benkendorf and the others tried to save as much of it as they could.

"First we hacked off the tusks and sent them aboard our boat. Then the natives tried to hew off the head, but this was slow work. As the belly of the brute was cut open, out rolled the intestines, and the stench was so dreadful that I could not avert my nausea and had to turn away. The stomach was well filled. Its contents were instructive and well preserved. They were chiefly young shoots of fir and pine. A quantity of young fir cones, also in a chewed state, were mixed with the mass."

Benkendorf's extraordinary find brought him face to face with one of the great mysteries of the Ice Age: What could have caused the sudden, worldwide extinction of the giant mammals, or megafauna, that inhabitated the Earth until only a few thousand years ago? After making a systematic review of past and present species, the zoologist Alfred Russel Wallace concluded several decades later:

"We live in a zoologically impoverished world, from which all the hugest, and fiercest, and strangest forms have recently disappeared. It is surely a marvellous fact, and one that has hardly been sufficiently dwelt upon, this sudden dying out of so many large mammalia, not in one place only but over half the land surfaces of the globe."*

The toll of death was especially apparent in Siberia, where according to a 19th Century geologist, "the bones of elephants are to be found in heaps along the shores of the icy seas from Archangel to the Bering Straits,

* The favoured explanation today is that these vast extinctions were not caused by any climatic changes but by primeval human hunters — despite their puny weapons. The evidence of many carcases suggest that in a typical Ice Age hunt they would lure thousands of animals into forest fires or drive them over cliffs.

forming whole islands of bones and mud at the mouths of rivers, and encased in icebergs, from which they are melted out by the heat of the short summer in sufficient numbers to form an important item of commerce."

More than 2,000 years ago, Chinese merchants were buying Siberian mammoth ivory for bowls, combs, knife handles and ornaments. From the descriptions of the natives, who had a superstitious dread of the great beasts that emerged each summer from their icy caskets, the Chinese gained the impression that the interior of the Earth was the natural residence of mammoths. A book attributed to the 17th century Chinese Emperor K'ang-hsi described the mammoth as "the underground rat of the north," which burrows through the ground with its huge teeth and "dies as soon as it comes into the air or is reached by sunlight."

Scott's Last Journey

The race to the South Pole in 1911 between the Norwegian Roald Amundsen and the Englishman Robert Falcon Scott was unfortunately little more than that — a race. Amundsen appears to have had no ambition but to reach the Pole before Scott.

Amundsen was tough, aggressive and secretive about his plans. Only when his fully prepared expedition was on its way to the Antarctic in 1909 did he send a challenging message to Scott, who was then in New Zealand: 'Heading south. Amundsen.' Scott followed him with perhaps too little preparation.

When Scott reached the Pole on 18 January 1912 he knew he had lost the race. He found Amundsen's abandoned tent with a boastful letter which he was asked to forward to the Norwegian King. There was nothing for it but to return 350 miles to his base at McMurdo Sound. His expedition was now in a pitiable state. His animals had died, and he had only four men with him: Dr E.A. Wilson, Lieutenant Henry Bowers, Captain Lawrence Oates and Petty Officer Edgar Evans. Willy Ley, in his outstanding book *The Poles*, tells the story of Scott's last journey:

As Scott's party started back, their food rations were low. They had established depots along the route, and his diary became a harrowing account, during the next two months, of struggling from one cache to another for the food to keep going.

Coming back down the glacier from the summit, Evans took a bad fall. Their pace slowed. On February 7, Scott recorded his "First Panic." One of their biscuit boxes was found to have a shortage — amounting to a full day's rations for the men. They were so close to starvation now that a single ration box was a vital matter.*

By mid-February, Evans showed signs of mental breakdown. On the 17th, he collapsed, fell into a coma and died. "Providence to our aid!" wrote Scott. "We can expect little from man now." Oates began to suffer from frost-bite as the temperatures went lower, day by day. He suffered intense pain for three weeks. One morning he asked the others what they thought he should do. "Nothing could be said," wrote Scott, "but to urge him to march as long as he could." On March 17 a blizzard blew. As the men huddled in their tent, Oates pulled himself heavily to his feet and said: "I am going outside and may be some time." With that, the stalwart Guards officer hobbled off into the whiteness. He was never seen again.

"It was the act of a brave man and an English gentlemen," wrote Scott. But it was already too late. The three survivors were so badly frost-bitten that they could hardly move. "Amputation is the least I can hope for," wrote Scott on March 19. "The weather doesn't give us a chance."

Day by day the marches had been getting shorter, until four and a half miles was as much as the men could move their sledge in a day. On the 21st, with just two days' rations left, and only 11 miles from a food depot, the terrible Antarctic winter closed in. For a week the three men were imprisoned by a fierce blizzard and no entries were made in the diary. Then, on March 29, Scott wrote:

"Every day we have been ready to start for our depot 11 *miles* away [his italics], but outside the door of the tent it remains a scene of whirling drift. I do not think we can hope for any better things now. We shall stick it out to the end, but we are getting weaker, of course, and the end cannot be far. It seems a pity, but I do not think I can write any more."

He signed his name: "R. Scott." The fading hand had strength for one last entry: "For God's sake, look after our people."

A search party found the tent and the bodies eight months later. Each man lay in his sleeping bag. Beneath Scott's arm were his diary and a few last letters. "Had we lived," Scott wrote in one of them, "I should have had a tale to tell of the hardihood, endurance and courage of my companions which would have stirred the hearts of every Englishman." The diary of Captain Scott serves as his monument, one of the imperishable documents

---

* The plentifully supplied Amundsen declared during *his* return from the South Pole: 'We had such masses of biscuits that we could positively throw them about.'

in the annals of exploration. Scott and his companions were not men to let polar exploration down. Even while struggling along and literally dying on their feet, they still carried with them 35 pounds of precious fossils and other geologic specimens collected *en route* to the South Pole — specimens which helped scientists to determine the age and history of that part of the Antarctic continent.

The First Free Diver

For thousands of years, people could dive down to 100 feet beneath the sea with the power of their unaided lungs — but they were never able to remain there for more than a few seconds.* At last, in 1819, Augustus Siebe invented a diving suit that received air pumped down from the surface. It enabled divers to stay for hours on the sea bottom. But they had to walk rather than swim. They moved in great discomfort because of the cumbrous weight of their apparatus, and they always depended on the assistance of others. Jacques-Yves Cousteau describes here his maiden dive with the compressed-air aqualung, which allowed the diver to stay two hours at a time at depths of 300 feet without needing air pipes, and swimming as naturally as a fish.

There was no petrol for civilians during the Axis occupation of France in the Second World War, and Cousteau's friend Émile Gagnan had invented a device for pumping cooking gas into car engines. It served just as well as the demand valve for an aqualung: it gave the diver air only when he wanted it. For the rest, Cousteau persuaded a naval friend to transform a gas-mask cylinder of soda lime, a small oxygen bottle, and a length of motor cycle inner tube into an artificial metal lung that repurified exhalations of air by filtering out carbon dioxide into the soda lime.

This is his account of how he first used the aqualung, from his book *The Silent World*.

---

* Aristotle said he had heard of divers being supplied by air from the surface by long tubes, and the first Emperor Claudius wrote of a people who "lived at the bottom of a lake". But these stories are not nowadays believed.

MENFISH

One morning in June 1943, I went to the railway station at Bandol on the French Riviera to collect a wooden case expressed from Paris. In it was a new and promising device, the result of years of struggle and dreams: an automatic compressed air diving lung conceived by Emile Gagnan and myself. I rushed it to Villa Barry where my diving comrades Philippe Tailliez and Frédérick Dumas were waiting. No children ever opened a Christmas present with more excitement than we did when we unpacked the first "aqualung." If it worked, diving could be revolutionized.

We found an assembly of three moderate-sized cylinders of compressed air, linked to an air regulator the size of an alarm clock. The bottles contained air condensed to one hundred and fifty times atmospheric pressure. From the regulator – there extended two tubes, joining on to a mouthpiece. With this equipment harnessed to the back, a watertight glass mask over the eyes and nose, and rubber foot fins, we intended to make unencumbered flights in the depths of the sea.

We hurried to a sheltered cove which would conceal our activities from curious bathers and Italian occupation troops. It was difficult to contain my excitement and discuss calmly the plan of the first dive. Dumas, the best goggle diver in France, would stay on shore keeping warm and rested, ready, if necessary, to dive to my aid. My wife Simone would swim out on the surface with a schnorkel breathing-tube and watch me through her submerged mask. If she signalled anything had gone wrong, Dumas could dive to me in seconds. "Didi," as he was known on the Riviera, could "skin dive" to sixty feet.

Staggering under the fifty-pound apparatus, I walked with a Charlie Chaplin waddle into the sea. A modest canyon opened below, full of dark green weeds, black sea urchins and small, flowerlike white algae. The sand sloped down into a clear blue infinity. The sun struck so brightly I had to squint. My arms hanging at my sides, I kicked the fins languidly and travelled down, gaining speed, watching the beach reeling past. I stopped kicking, and the momentum carried me on a fabulous glide. When I stopped, I slowly emptied my lungs and held my breath. The diminished volume of my body decreased the lifting force of water, and I sank dreamily down. I inhaled a great chestful and retained it. I rose towards the surface.

I reached the bottom in a state of excitement. A school of goat bream, round and flat as saucers, swam in a rocky chaos. I looked up and saw the surface shining like a defective mirror. In the centre of the looking-glass was the trim silhouette of Simone, reduced to the size of a doll. I waved: the doll waved back.

I swam across the rocks and compared myself favourably with the fish. To swim fishlike, horizontally, was the logical method in a medium eight hundred times denser than air. To halt and hang attached to nothing, no lines or pipes to the surface, was a dream. I had often had visions of flying by extending my arms as wings. Now I flew without wings.

I thought of the helmet diver arriving on his ponderous boots at such a depth as this and struggling to walk a few yards, obsessed with his umbilical cords and his head imprisoned in copper. I had seen him leaning dangerously forward to make a step, clamped in heavier pressure at the ankles than the head, a cripple in an alien land. From this day forward we would swim across miles of country no man had ever known, free and level, with our flesh feeling what the fish scales know.

2
Air
and Space

Too Close to the Sun

The story of Icarus, supposedly the first aviator, who died because the Sun melted the wax that secured his feathered wings, although obviously impossible, is a dream of future flight that makes a powerful tragic tale. It is told here by the Roman poet Ovid.

Daedalus appears to have been a real historical character. He was the inventor who designed the labyrinth in the king's palace in Crete in which Theseus stalked the Minotaur. His very name in Greek came to mean 'cunningly wrought' and still means 'maze' in French. At the time of this story he was imprisoned in Crete by Minos its king, who feared the loss of his talents. This episode seems characteristic of him.

Daedalus, tired of Crete and of his own absence from home, was filled with longing for his own country, since he was shut in by the sea. Then he said: 'The king may block my way by land or across the ocean, but the sky, surely, is open, and that is how we shall go. Minos may possess all the rest, but he does not possess the air.'

With these words, he set his mind to sciences never explored before, and altered the laws of nature. He laid down a row of feathers, beginning with tiny ones and gradually increasing their length, so that the edges sloped upwards. He made them in the fashion of shepherds' pipes that are built up from reeds, each slightly longer than the last. Then he fastened the feathers together in the middle with thread, and at the bottom with wax; when he had arranged them in this way he bent them round into a gentle curve, to look like real birds' wings. His son Icarus stood beside him and, not knowing that the materials his father was handling were to endanger his own life, he laughingly captured the feathers that blew away in the wind, or softened the yellow wax with his thumb, and by his pranks hindered the marvellous work in which his father was engaged.

When Daedalus had put the finishing touches to his invention, he raised himself into the air, balancing his body on his two wings, and there he hovered, moving his feathers up and down.

Then he prepared his son to fly too. 'I warn you, Icarus,' he said, 'you must follow a course midway between earth and heaven, in case the Sun should scorch your feathers if you go too high, or the water make them heavy if you are too low. Fly halfway between the two. And pay no attention to the stars, to Boötes, or Helice or Orion with his drawn sword: now take me as your guide and follow me!'

While he was giving Icarus these instructions on how to fly, Daedalus was at the same time fastening the novel wings on his son's shoulders. As he worked and talked the old man's cheeks were wet with tears, and his fatherly affection made his hands tremble. He kissed his son, whom he was never to kiss again: then, raising himself on his wings, flew in front, showing anxious concern for his companion, just like a bird who has brought her tender fledglings out of their nest in the treetops, and launched them into the air. He urged Icarus to follow close, and instructed his son in the art that was to be his ruin, moving his own wings and keeping a watchful eye on those of Icarus behind him. Some fisherman, perhaps, plying his quivering rod, some shepherd leaning on his staff, or a peasant bent over his plough handle caught sight of them as they flew past and stood stock still in astonishment, believing that these creatures who could fly through the air must be gods.

Now Juno's sacred isle of Samos lay on the left, Delos and Paxos were already behind them, and Lebinthus was on their right hand, along with Calymne, rich in honey, when the boy Icarus began to enjoy the thrill of swooping boldly through the air. Drawn on by his eagerness for the open sky, he left his guide and soared upwards, till he came too close to the blazing Sun, and it softened the sweet-smelling wax that bound his wings together.

The wax melted. Icarus moved his bare arms up and down, but without their feathers he had no purchase on the air. Even as his lips were crying his father's name, they were swallowed up in the deep blue waters that are called after him. The unhappy father, a father no longer, cried out: 'Icarus! Icarus! Where Are you? Where am I to look for you?' As he was still calling 'Icarus' he saw the feathers on the water, and cursed his inventive skill. He laid his son to rest in a tomb, and the land took its name from that of the boy who was buried there [the island of Icaria].

'The Dangerous Deserts of Space'

Dr Peter Howard, of the RAF Institute of Aviation Medicine at Farnborough, describes a tragic early flight that warned of the dangers of the air.

In 1875, three early 'astronauts,' Gaston Tissandier, Joseph Croce-Spinelli and Theodore Sivel made an ascent in a balloon, the *Zenith,* that was surely a landmark in the exploration of space. Only Tissandier survived, and the deaths of his companions raised physiological questions vital to aviation.

The Société de Navigation Aérienne of France had sponsored an ascent to great height. The balloon's scientific apparatus included eight barometers to prove the altitude reached. Most important were three small balloons containing air with 70 per cent of added oxygen, each connected to a mouthpiece. They were carried on the advice of Paul Bert, whose research into the effects of decreased barometric pressure is still the great classic of aviation physiology. During an ascent to a simulated altitude of 24,000 feet in Bert's laboratory, Croce-Spinelli and Sivel had both become convinced of the benefits of breathing oxygen. Not surprisingly, they felt they could go even higher with the help of this 'powerful cordial.'

After an uneventful ascent of the *Zenith* to 11,000 feet, gas began to escape from the open end of the envelope. The smell was strong, and Croce-Spinelli felt depressed. The coincidence that the three passengers all became unwell soon afterwards led many to believe that the escaping gas helped to kill Sivel and Croce-Spinelli.

By the time they reached 23,000 feet all three were showing signs of a lack of oxygen. Tissandier felt dejected, and disinclined to work the instruments. Sivel began to close his eyes for a few moments at a time, and had to force himself into activity. Croce-Spinelli was the least affected and continued to peer through his spectroscope. Having made a small discovery, his excitement was unnaturally great. His face was 'radiant with joy' and he 'earnestly entreated' Tissandier to read the instrument.

His reactions, in short, were out of proportion to the triviality of the observation he had made.

The only physiological data was taken at about 16,000 feet, incomplete records of pulse and respiratory rates. All three travellers had a fast pulse. Sivel's heart-rate was almost twice normal. Anoxia, lack of oxygen in the blood, was undoubtedly the main cause.

Tissandier found it difficult to write. His notes, which he had no clear recollection of making, were almost illegible, but he saw that his colleagues were drowsy and that Sivel's breathing was laboured. All three now began to use the oxygen mixture sporadically, which revived Sivel so much that he remembered his ambition to go yet higher. His face lit up, and he asked Tissandier for the barometer reading. It was 300 millimetres of mercury, equivalent to 24,500 feet. Sivel, impatient to go on, asked whether he should throw out ballast. The responses of his colleagues show how severely affected they were. Tissandier told him to please himself, while Croce-Spinelli nodded, unable to speak.

There was now little sign of the three enthusiastic adventurers who had set out so bravely. Three of the remaining five bags of ballast were emptied, and the balloon ascended rapidly. Tissandier soon grew so feeble that he could not even turn his head to look at his fellows. He knew he needed oxygen, but the effort of reaching for the mouthpiece was too much for him. He could see the barometer and its needle told him they had reached and exceeded the 26,000 feet they had all dreamed of. Tissandier could not open his mouth to tell his friends; it is doubtful whether they would have heard.

Suddenly, his eyes closed and he fell unconscious to the floor of the basket.

He awoke half an hour later to find that the balloon was falling rapidly. He jettisoned some ballast to slow the descent, and was able to write in the log that Sivel and Croce-Spinelli were lying inert.

Then Tissandier lost his senses once more. Croce-Spinelli came to and shook him back to awareness, although he himself was barely capable of action. He was just able to throw out the rest of the ballast, in an attempt to check the speed of the descent.

The later behaviour of the balloon is largely conjectural, for Tissandier soon collapsed once more.

It certainly climbed again and the barometers recorded the greatest altitude reached as 28,000 feet. Tissandier was insensible for about an hour and a half. When he awoke, the balloon was once more descending and at a frightening speed. Sivel and Croce-Spinelli lay at the bottom of the basket. Tissandier crawled over to rouse them. Their faces were black and their mouths filled with blood. Tissandier, in the state that he was in, did not realise they were dead. He tried wildly to slow the fall of the balloon, which was by now very near the ground. At last he succeeded in finding and freeing the anchor, and was able to break the shock of the impact as it hit the ground.

Climbing from the basket, Tissandier now realised the enormity of the situation and he behaved like one demented, throwing himself first against one corpse, then against the other, imploring them to speak.

A fortnight later, with scientific detachment, he published a complete account of the affair in the scientific journal *La Nature*, of which he was editor.

Croce-Spinelli, and probably Sivel, died during the second ascent of the balloon. Tissandier concluded that their deaths were due to anoxia. He suggested that it was possible to survive relatively short exposures to low barometric pressures, but that the stress of almost two hours of continuous exposure to high altitude cannot be borne.

He was rightly convinced that at the very time when his friends most needed oxygen, they were too feeble to reach their breathing tubes.

The first fatal event in the flight was the escape of gas from the envelope. The lighting gas used to inflate the balloon contained a high proportion of carbon monoxide. This gas has a great affinity for haemoglobin, and consequently interferes with the carriage of oxygen by the blood. It is therefore possible that all three members of the party were suffering from mild coal-gas poisoning, which would aggravate the effects of the later lack of oxygen.

The first unequivocal signs of anoxia apparently occurred at 23,000 feet, but it must be remembered that to recognise abnormal behaviour requires a normal observer. One of the most dangerous features of oxygen lack is loss of judgment; not only does the sufferer fail to recognise the peculiar actions of others, but he regards himself as unusually rational and clear-headed. This stage is usually reached above 15,000 feet, and had Tissandier's critical faculty not been impaired he would have surely noticed some unusual behaviour in the others. Significantly, his notes refer first to the fact that he had breathed oxygen, and then to the laboured breathing and drowsiness of Sivel and Croce-Spinelli. Without the help of the gas mixture he would probably have been unable to write at all, for most people rapidly became stuporous at this altitude and lose consciousness quite quickly.

The early feeling of elation and of power gives place to indifference and irritability. Croce-Spinelli's childish delight at the spectroscopic findings was certainly pathological, and had either of the others spoken a sharp word of disbelief or rebuke he would probably have burst into tears.

Croce-Spinelli recovered for a while when the balloon descended to about 22,000 feet. He at least had enough strength to shake his companion, and he clearly did not realise the danger of this action. Even the slightest exertion by someone suffering from anoxia is enough to tip the scale towards collapse. During the Second World War, for example, it was common for a crew member, helping an unconscious fellow, to fall comatose from the effort.

It might seem strange that deaths from anoxia should have occurred despite the oxygen mixture. It should have served them up to an altitude of 35,000 feet. Clearly, they did not perish because their supply of the gas was not rich enough.

There is no record of the capacity of the three small balloons that contained the oxygen. The volume of gas available probably did not exceed 150 litres for each voyager. Now even under conditions of rest, they would each have needed about 10 litres per minute. Taking into account the excitement of the occasion, and the work of manipulating instruments and moving about, even 20 litres a minute for each man is too little. In short, they each had only a six-minute supply of it for a flight of which nearly two hours and a half were spent above a height where it was essential. This seems the more incredible when it is remembered that Croce-Spinelli and Sivel were convinced of the need for extra oxygen by their experience at Paul Bert's laboratory.

They had apparently grasped the principle that anoxia could be prevented by a suitable gas mixture without appreciating the need for its continuous use. It is perhaps understandable that these two engineers should grasp the physics and ignore the physiology, but it is surprising that Paul Bert should have allowed the expedition to set out with a totally inadequate quantity of gas. The key to the whole tragedy lies in this mysterious oversight. At the time when the apparatus for the *Zenith* was being assembled, Bert was away from Paris, and his advice had to be sought by letter.

He recommended the 70 per cent oxygen mixture, but it was not until later that he learned of the small amount of it they were taking. Bert immediately wrote to Croce-Spinelli warning him of its pitiful inadequacy. But perhaps because of a natural reluctance to delay the flight, Croce-Spinelli and the others interpreted this as meaning that the oxygen mixtures should not be used until the last possible moment. It was, without doubt, the euphoria of anoxia that delayed the "last possible moment" until physical strength could no longer match mental resolution.

The final puzzle is Tissandier's survival. Theories to explain it have included his own suggestion that his more complete unconsciousness protected him. Yet this does not carry much conviction. The simple truth is that there is a very wide variation between individuals in their tolerance to lack of oxygen. A degree of anoxia that will strike one person senseless may leave another still capable of moving about, albeit rather drunkenly.

As the reachable altitudes have increased, so the development of oxygen equipment has improved until pressure garments or space-suits have superseded the simple face-mask. But, just as in Tissandier's day, the failure of this equipment is the thing most to be feared. The insidious onset of oxygen lack, and the loss of insight which it engenders, lead to a disregard of the danger signs. Death from anoxia will surely overtake not a few of those who were described, at the funeral of the *Zenith* victims, as "the others who will one day explore the dangerous deserts of space."

Faster than Sound

Flight faster than the speed of sound, about 670 mph, was long held to be fundamentally impossible by the same kind of 'experts' who had once maintained that powered flight itself was impossible.* They maintained that at or beyond the speed of sound, the force of wind would render an aircraft uncontrollable and force it to crash.

So much for theory. Charles 'Chuck' E. Yeager of the U.S Air Force was no theorist but one of the world's finest test pilots. On 14 October 1947, over Edwards Air Force Base, California, he broke the sound barrier in the Bell XS-1 rocket plane *Glamorous Glennis* that he had named after his wife. Here he laconically describes his flight as he separated from the B-29 bomber which had carried him aloft:

Bob Cardenas, the B-29 driver, asked if I was ready.

"Hell, yes," I said. "Let's get it over with."

He dropped the X-1 at 20,000 feet, but his dive speed was once again too slow and the X-1 started to stall. I fought it with the control wheel for about five hundred feet, and finally got her nose down. The moment we picked up speed I fired all four rocket chambers in rapid sequence. We climbed at .88 Mach** and began to buffet, so I flipped the stabiliser switch and changed

---

* By a curious coincidence, the speed of sound on Earth is almost exactly a millionth of the speed of light. But it *can only be* a coincidence, since on other planets, with different atmospheric densities, the speed of sound would be different. The speed of light, at 670 million mph, is of course constant everywhere in the Universe, as far as we know.

** Mach 1 was equal to the speed of sound. Mach 2 would be twice this speed, and so forth. (The Earth's 'escape velocity,' the speed which a plane or rocket must attain to reach orbit and "escape" from the planet is 25,000 mph or Mach 37.) These numbers are named after the great Austrian physicist Ernst Mach, famous for his experiments with airflow.

the setting two degrees. We smoothed right out, and at 36,000 feet I turned off two rocket chambers. At 40,000 feet, I had thirty per cent of my fuel left, so I turned on rocket chamber three and immediately reached .96 Mach. I noticed that the faster I got, the smoother the ride.

Suddenly the Mach needle began to fluctuate. It went up to .965 Mach — then tipped right off the scale. I thought I was seeing things! We were flying supersonic! And it was as smooth as a baby's bottom: Grandma could be sitting up there sipping lemonade. I kept the speed of the scale for about twenty seconds, then I raised the nose to slow down.

I was thunderstruck. After all the anxiety, breaking the sound barrier turned out to be a perfectly paved speedway. I radioed Bob in the B-29.

"Hey, that Machmeter is acting screwy. It just went off the scale on me."

"Fluctuated off?"

"Yeah, at point nine-six-five."

"Son, you is imagining things."

"Must be. I'm still wearing my ears, and nothing else fell off, neither."

The guys in the NACA* tracking van interrupted to report that they heard what sounded like a distant rumble of thunder: my sonic boom! The first one by an airplane ever heard on Earth. The X-1 was supposedly capable of reaching twice the speed of sound, but the Machmeter aboard only registered to 1.0 Mach, which showed how much confidence they had.

'Call Me Doctor!'

The Prussian-born rocket pioneer Wernher von Braun could be a most prickly person, as the journalist Oriana Fallaci discovered during an interview:

Bart Slattery, Wernher von Braun's public relations officer, was all humble and obedient, his face lifeless and anxious. He looked at von Braun as a pupil looks at its master and seemed to be asking: In what way can I serve you, my lord? He could serve him by reminding me that

---

* The National Advisory Committee on Aeronautics, the predessor to the space agency NASA.

Wernher von Braun had no time to lose, that he could spare half an hour at the most.

Slattery looked at his stopwatch to indicate that the half hour began as from that minute. I took a deep breath and went into the attack.

"Mr von Braun, I'll omit the preliminaries and put a question to you at once. The question is this . . ."

I was interrupted by a white slip of paper that came sliding across the table in front of me. It came from Slattery, who had written on it: "*Doctor* von Braun. *Not* Mr von Braun." The first word was written with angry force and in big capital letters: DOCTOR. I stared at him in stupefaction, my face flaming, and threw a quick glance at von Braun in the secret hope that he would call Slattery a fool. But von Braun was busy examining a fingernail, like someone who has noticed nothing.

I began again.

"The question is this, Mr von Braun . . ."

Another white slip of paper came from Slattery. Enraged. "DOCTOR von Braun!!!" Von Braun was still looking at his fingernail. Goodness knows what was the matter with it.

"Will the journey to the Moon really take place by 1970, *Doctor* von Braun?"

Slattery nodded happily. Von Braun quit scrutinising his fingernail and began to answer my question.

'We Came In Peace For All Mankind'

One of the greatest moments in history was man's first footfall on another world. This dispatch of mine from Houston described the culminating moment of Apollo 11, America's first moon-landing mission. It appeared in *The Daily Telegraph* on 22 July 1969:

The two American lunarnauts, Neil Armstrong, 38, and Colonel Edwin "Buzz" Aldrin, 39, were leaping about on the Moon's surface like children yesterday morning when there came a surprise interruption from President Nixon.

They were discovering that H.G. Wells was correct in his novel "The First Men on the Moon" in speculating that a man could move rapidly across Moon gravity — which is one sixth that of Earth's — with little hops and jumps.

Suddenly, in the midst of these gambolings, there came a startling interruption. Armstrong and Aldrin came to an abrupt halt.

The announcement came over their headsets: "The President would like to speak to you." The two men stood still and assumed a dignified demeanour.

"Hello, Neil and Buzz," came Mr. Nixon's very loud and clear voice as he made what he called "the most historic telephone call ever made."

"I just can't tell you how proud we all are of you," he went on. "Because of what you have done, the heavens have become part of man's world. And as you talk to us from the Sea of Tranquillity, it inspires us to double our efforts to bring peace and tranquillity to Earth.

"For one priceless moment in the whole history of man, all the people on earth are truly one. One in their pride in what you have done. And one in our prayers that you will return safely to Earth."

Armstrong hastily summoned his thoughts to reply.

"Thank you, Mr. President. It is a great honour and privilege for us to be here representing not only the United States but men of peace of all nations. And with interest and curiosity and a vision for the future. It's an honour for us to be able to participate here today."

Earlier, the progress of the two men's emergence from the lunar landing craft Eagle had been highlighted by vivid comments among radio exchanges with Mission Control. "The hatch is coming open," we heard Armstrong say.

Then, twelve minutes later, he was on the porch at the top of the ladder. Aldrin was saying to him: "Neil, you're lined up nicely. Towards me a little bit. Okay, down. Over here, roll to the left. Put your left foot to the right a little bit. Okay, that's good. Roll left."

"How am I doing?" asked Armstrong.

"You're doing fine," said Aldrin.

"Now do you want those bags [for collecting rock samples]?" asked Aldrin.

"Yeah, got it," said Armstrong.

By this time, he had begun to operate the television camera.

Mission Control interrupted: "Man, we're beginning to get a picture on the TV."

(At this point, Armstrong's foot could be seen at the top of the spacecraft's ladder on the TV screens at Houston. Thousands of journalists clapped and cheered.)

"Okay, Neil," said Mission Control. "We can see you coming down the ladder."

It was a few moments later. "I'm at the bottom of the ladder," Armstrong called out. "The spacecraft footpads are only depressed in the surface about one or two inches, although the surface appears to be very, very fine-grained as you get close to it. It's almost like a powder."*

Then he uttered the sentence which rang round the world and will surely appear in dictionaries of quotations to be published centuries from now:

"I'm going to step off the spacecraft now. That's one small step for a man. One giant step for mankind."

"The surface is fine and powdery," he continued. "I can pick it up loosely with my toe. It does adhere in fine layers like powdered charcoal to the sole and side of my boots. I only go in a small fraction of an inch, but I can see the footprints of my boots and the treads in the fine sandy particles.

"It's actually no trouble to walk around. We're on a very level place here. I can see some evidence of rays [escaping gas] emanating from the descent engine of the spacecraft, but in a very insignificant amount."

Aldrin called out from Eagle: "It looks beautiful from here, Neil."

"It [the landscape] has a stark beauty all its own," came Armstrong's reply. "It's like much of the high desert of the United States." Then he reported on the progress of his work: "Contingency sample is in the pocket."

Aldrin descended successfully, and the two men practised walking about. Aldrin found the walking somewhat peculiar.

"Got to be careful that you are leaning in the direction you want to go. In other words, you have to cross your foot over for it to stay underneath where your centre of mass is."

Then they unveiled the plaque that will commemorate their visit for millions of years — until it is corroded beyond recognition by microme-teorites. It bore a picture of the Earth's two hemispheres and read: "Here man from the planet Earth first set foot upon the Moon, July, 1969 AD. We came in peace for all mankind." It carried the signatures of Armstrong, Aldrin and President Nixon.

Now they were putting the television camera in position on its tripod for a panoramic view. The Moon's surface was revealed to people on earth in great clarity. Armstrong described the landscape: "The little hill beyond the shadows of the spacecraft is a pair of elongated craters. [The rims of the

---

* Confirmation that the Moon had a hard surface, although made earlier by unmanned spacecraft, came as a vast relief to would-be lunar colonists. Professor Thomas Gold and others had created considerable alarm by predicting that the Moon's sandy surface was *soft*, and that people and spaceships would vanish into it, like stones dropping into the sea. This idea formed the plot of Arthur C. Clarke's terrifying 1961 novel *A Fall of Moondust*.

craters made it look like a hill.] They are forty feet long and twenty feet across and probably six feet deep."

Finally there came an ominous warning from Houston.

"You've got about ten minutes left now prior to commencing your EVA termination activities." In other words, they must start preparing to re-embark in Eagle: otherwise their oxygen could run out. [EVA, in the tortured NASA prose, meant "extra vehicular activity."

Aldrin presently announced: "The hatch is closed and latched." Michael Collins [in orbit above them] heard the news and said: 'Hallelujah."

'We've had a Problem'

Journeys to the Moon did not always go so smoothly. The crew of Apollo 13 had to fight for their lives when part of their spacecraft blew up nearly a quarter of a million miles from Earth. Peter Bond gives an exciting account of this episode.

Captain James Lovell, 42, commander of the Apollo 13 Moon-landing mission, had been an astronaut for nearly eight years. His lunar module pilot was Fred Haise, who became an astronaut in 1966, at the same time as the command module pilot, Jack Swigert, 38.

Apollo 13 set off from Cape Kennedy on April 11 1970. The message from ground control was: "Good Luck. Head for the hills," a reference to their lunar destination of the Fra Mauro uplands.

The spacecraft performed perfectly as it was boosted out of Earth orbit. The command service module (CSM), Odyssey, separated from the third stage and the lunar module (LM), named Aquarius, was successfully docked.

The ship drifted along on her "free-return" course until the evening of the second day when a short engine burn sent her towards lunar orbit. The flight controllers were confident as the flight entered its third day.

But there were minor problems which later assumed a much greater significance. Swigert had been having trouble reading the gauge for one of the oxygen tanks. It had gone off the scale, causing ground control to make frequent requests to stir up the oxygen. There were also worries over helium pressure in the lunar module descent stage. And Lovell reported low pressure in one of the hydrogen tanks. "Just after we went to sleep last night we had a master alarm and it really scared us. We were all over the cockpit like a wet noodle."

At 55 hours into the flight, the crew were making another of their TV broadcasts as they opened up Aquarius for the first time.

"What we plan to do today," Lovell began, "is start out in the spaceship Odyssey and take you on through the tunnel into Aquarius." Lovell's camera followed the white, ghostly figure of the tour guide Haise through the tunnel into the LM. There he demonstrated some of the equipment to be used on the Moon.

Ten minutes after they went off the air, the astronauts' hopes for a pleasant evening were suddenly shattered. Lovell said later: "All three of us heard a rather large bang."

It was 9.11 pm at Mission Control in Houston. The crisis had begun five minutes earlier, unknown to them or the crew. An amber warning light had flashed on, indicating low pressure once more in a hydrogen tank in the service module. As before, a message was transmitted to Swigert:

"13, we've got one more item for you when you get a chance. We'd like you to stir up the oxygen."

The CM [command module] pilot threw the four switches that would set the fans in motion in the hydrogen and oxygen tanks. In the CSM, attached to the rear of the command module, the wires in oxygen tank 2 were almost bare of insulation — later investigations blamed overheating during the prolonged emptying of the oxygen tank more than two weeks earlier. Sixteen seconds later, an arc of electricity jumped across two wires, causing a fire in the oxygen tank and a rapid increase in oxygen pressure. This went unnoticed for some time since the hydrogen pressure warning light overrode the warning light system for oxygen pressure. Leaking oxygen now spread the fire through the whole interior of Bay 4 in the CSM. The oxygen, together with gases from burning insulation, finally blew out the weakest part of the bay — the panel on the craft's outer hull — with a loud bang. Swigert then heard a master alarm in his earphones, and an amber warning light signalled a power drop in main bus B. He slammed the hatch shut behind Haise as he emerged from the tunnel, and returned to his seat to inform Mission Control:

"OK, Houston, we've had a problem."

Jack Lousma of Mission Control was unprepared for the sudden announcement.

"Say again please."

"We've had a problem," Swigert repeated. "We've had a main bus B undervolt."

"And a pretty large bang," Haise added.

The urgency in their voices made the flight controllers sit up. No one knew what had happened. The telemetry from the craft could not reveal that an oxygen tank had blown up. It could only send back a confusing pattern of pressure and temperature readings. The reason for the loss of power was not immediately obvious either. In fact, the explosion had closed oxygen supply line valves to fuel cells 1 and 3 which had then ceased to operate. Without all three fuel cells working functioning normally, the Moon landing was automatically cancelled. The seriousness of the position dawned as Lovell reported that the oxygen pressure gauge for tank 2 was now reading zero and the pressure in tank 1 was noticeably dropping. Oxygen in the fuel cells was vital both for electricity and was the source of nearly all water and air. It began to look increasingly likely that the crew would die from suffocation in a dark, cold spacecraft drifting out of control in the vast emptiness of space.

Lovell said later: "I looked out of the window and saw this venting. My concern went from: 'I wonder what this is going to do to the landing,' to: 'I wonder if we can get back home again.' When I saw both oxygen pressures, one actually at zero and the other one going down, it dawned on me that we were in serious trouble."

He floated out of his seat to peer through the window. He saw a white, wispy cloud surrounding the CSM. "We are venting something into space," he told a disturbed Jack Lousma.

This explained why the spacecraft kept veering off course. The thrust from the leaking gas was counteracting Lovell's manual manoeuvring. It was also threatening to deprive Apollo 13 of its most important navigational aid.

Chief flight director Gene Kranz set enquiries in hand.

"OK, let's everybody think of the kind of things we'd be venting. GNC [guidance and navigation control officer], you got anything that looks that normal in your system?"

The reply was negative.

"OK," said Kranz, "let's everybody keep cool. We've got the LM still attached."

It soon became clear that the presence of the LM was the one saving grace in a potentially fatal crisis, for it, alone, was unaffected by whatever was disabling the CSM.

The situation in Odyssey continued to deteriorate. Communications with the ground were poor, and sometimes ceased altogether as the craft continued to wobble. Two fuel cells were dead, and one oxygen tank was registering zero. The CSM's electricity would last only as long as oxygen continued to reach it from the other tank, but the power output from the surviving fuel cell was steadily dropping, as was the oxygen in the second tank. Kranz ordered the astronauts to begin powering down the CM to reduce the strain on the main bus A.

The astronauts had been boosting the power output from main bus A by connecting it to the re-entry battery, a back-up supply designed to last for up to ten hours. But as the power drain continued, they were told to disconnect it; the battery would be their sole source of electricity during re-entry, and without it they were dead. A further precaution followed when the oxygen surge tank in the CM was isolated, so ensuring oxygen supply during re-entry. Mission Control then closed the valves between fuel cell 3 and the oxygen tanks in the vain hope that it was the fuel cell, not the tanks, which was leaking. The closure made no difference to the fall in oxygen pressure, and it was at last clear to everyone that the LM would have to act as a life raft for the next four days.

New problems immediately arose. Although some tests of the LM had been carried out in Earth orbit by the crew of Apollo 9, there had never been any practice of a whole Apollo crew relying solely on the LM. More significantly, the LM was designed to keep two men alive for up to 50 hours, whereas it was now required to sustain three men for 84 hours. All course corrections would have to be made by the LM descent engine, including the vital burn that would return Apollo 13 to an Earth rendezvous trajectory.

With the CM shut down completely, and the crew relying totally on the LM for survival, the question now arose about the safest way to get home. The idea of simply reversing course was soon squashed; the LM's descent engine had no power for such a manoeuvre, and a failure would make the craft crash into the Moon. Yet assuming a successful burn to place Apollo in a free-return trajectory, the time to re-entry would have been an unacceptable 100 hours. Ways would have to be found to stretch the air and provisions while reducing the flight time. Fortunately, it soon became clear that there was enough oxygen to last the four days, though electricity and water were very much on the borderline.

The flight controllers at last agreed to two burns, the first to place the head of Apollo towards Earth, and the second after the craft passed behind the Moon to boost it on its way. At 61 and a half hours into the mission, and nearly six hours after the bang, Lovell fired the LM descent engine manually for 30 seconds. The first step on the long road home had been successfully taken.

There would be little sleep for the flight controllers that night as they worried about the rest of the flight. The astronauts had a chance to rest. There were no seats in the LM, with standing room only for two men, so while two of them remained in Aquarius, the third moved into Odyssey and tried to sleep. The rota began at around 4 am when Lovell sent Haise to his "bed" on the couch in the dark, silent command module. Fatigue and stress were affecting all three by now, so that Lovell found it difficult to control the craft's attitude by operating the LM's thrusters manually. His task was made even more difficult since the LM was positioned at one end of the combined craft and its thrusters were not designed to control a spacecraft several times heavier than itself. Furthermore, the dashboard display for the guidance platform had been switched off to conserve power. It was hardly surprising that he sometimes forgot which direction the craft was drifting to, or that he once turned the craft completely around.

Lovell finally placed Apollo in the correct attitude — sideways on to the flight path. To prevent overheating by the Sun, he rotated Apollo through 90 degrees once every hour.

When the morning came, leading NASA executives debated the type of burn Apollo 13 would make that evening, two hours after it had passed behind the Moon. The scheme eventually adopted was that the CSM would remain attached and a Pacific splashdown would occur.

The crew were too busy to notice how swiftly they were approaching the Moon. When Apollo entered the Moon's shadow, Lovell had already put the craft into position for the burn, so they could relax a little as they passed out of contact with Mission Control and swept to within 160 miles of the lunar surface. Eight minutes after losing radio contact, the Apollo crew watched the Sun rise above the lunar horizon, enabling Haise and Swigert to grab their cameras for souvenirs of the world on which they were destined never to set foot. After 25 minutes of radio silence, they reappeared around the limb of the Moon. They were on their way home at last.

At 79 hours into the mission, Lovell manually fired the LM descent engine for a total of 4 minutes 23 seconds, gradually increasing the thrust.

But elation turned to dismay on the ground as tracking stations showed that Apollo 13 was moving further and further off course. Something seemed to be venting. The astronauts were not immediately told. They had been psychologically boosted by the successful burn, but they were tired and increasingly irritable.

Capcom Jack Lousma suggested that the astronauts might wear their spacesuits to combat the falling temperature — now below 10 deg.C — but they declined, feeling that they would be too cumbersome. There were blankets in their emergency landing pack, but these were left buried under a mound of equipment in the command module.

The astronauts spent the afternoon of April 15 preparing for yet another crucial engine burn to correct the drift off course. If it failed, the craft would skip off the Earth's upper atmosphere and be lost for ever. At 10.31 pm, the LM descent engine was fired yet again for 15 seconds, a burst intended to slow the craft by just 7 feet per second so that it would hit the narrow re-entry corridor through the atmosphere.

Lovell later recalled his feelings:

"When the ground read out the procedure to us, I just couldn't believe it. I thought I'd never have to use something as way out as this. Because it was a manual burn, we had a three-man operation. Jack would tell us when to fire the engine, and when to stop it. Fred handled the roll manoeuvre and pushed the buttons to start and stop the engine."

Lovell himself had to keep the telescope cross hair just touching the "horns" of the crescent Earth by manually firing the altitude thrusters. The LM was powered down again so that it used only 10–12 amps of current. Lovell stayed on watch while the other two rested.

The helium tank blew off its surplus pressure at about 2 am on April 16. Lovell then had to struggle to set up the thermal roll once more. He was so bleary-eyed that at one stage he was confused over whether the Earth or the Moon was passing across the window.

The doctors were worried about their fatigue. As they neared Earth, the rationing forced by the freezing up of the command module water tank meant that they were drinking only 6 ounces a day. This caused growing concern lest dehydration provoked errors.

The last 24 hours were the longest that anyone had experienced. Lovell, like everyone else, was thinking about the re-entry. At about 6.30 pm Capcom Vance Brand told him that the re-entry checklist was ready at last, starting with Swigert's command module, their only means of returning safely to Earth.

"He'll need a lot of paper," warned Brand.

There followed a series of delays over a lack of copies for the flight engineers to follow, causing Lovell to cut in sharply: "We just can't wait around here to read the procedures all the time up to the burn! We've got to get them up here, look at them, and then we've got to sleep!" At last, from the ground, astronaut Thomas Mattingly read up the checklist, a task that took nearly two hours.

By now the first and second batteries on the CM had been fully charged by Aquarius. Its descent stage ran out of water soon after, but the crew were able to switch over to the ascent stage supply. Mission Control demanded a minor course adjustment about five hours before re-entry, to make sure that Apollo hit the entry corridor — there was a fear that separation from the LM might push Odyssey off course.

The early hours of April 17 showed more signs of stress among the shipwrecked crew. Jack Lousma was reading up a number of checklist alterations to Haise and Swigert when Lovell interrupted:

"OK, Jack, this is Jim. I just want to make sure that any of the changes to the checklist that come up, you make sure that they're absolutely essential."

Lousma radioed up to the shivering astronauts:

"I wish we could figure out a way to get a hot cup of coffee up to you. It would taste pretty good now wouldn't it?"

Lovell sounded miserable:

"Yes, it sure would. You don't realise how cold this thing has become. The Sun is simply turning on the engine of the service module. It's not getting down to the spacecraft at all."

But Lousma was soon able to give them some welcome news.

"OK, skipper, we figured out a way for you to keep warm. We decided to start powering you up now."

With nine and a half hours still to go before splashdown, Lovell wanted reassurance:

"Sounds good. And you're sure we have plenty of electrical power to do this?"

Lousma replied in the affirmative. The temperature in Aquarius crept up, eventually enabling the crew to switch on the window heaters. By 5 am they were baking in a cabin temperature of 16 deg C. Even Odyssey was warming up, as Swigert commented:

"It's almost comfortable." Much to his relief, main bus B, the first to fail after the explosion, was found to work perfectly. He began to defrost the CM's thrusters.

Five hours before splashdown the LM thrusters were fired for 23 seconds to increase Apollo's speed by 3 feet a second and to ensure an accurate splashdown near the recovery ships. Swigert, having had to rely on his companions during most of the flight, was happy now to be working in Odyssey once more. The most difficult task was aligning the CM's guidance platform. The cloud of "little white fluffy objects" surrounding the craft obliterated the stars when he sought them in his sextant. He eventually succeeded in lining up the stars Altair and Vega on the shaded side of the craft with the light off.

Lovell now manoeuvred Apollo to the attitude needed for safe separation from Aquarius. Switching to automatic control, he told Capcom that he was going to bail out of the LM. Before closing the hatch, he looked back at their life raft. It looked like "a packed garbage can." Safe inside the CM, the astronauts prepared to jettison Aquarius. The explosive charges were fired 90 minutes before re-entry to send the LM on its final journey. Mission Control radioed:

"Farewell, Aquarius, and we thank you."

"She was a good ship," Lovell added. Haise kept taking photographs until Aquarius was a mere speck of light.

Mission Control seemed happy with the trajectory as the moment of truth approached.

Nine minutes later, Odyssey skipped into the upper atmosphere, subjecting the crew to a welcome of more than 5G. The heat shield withstood the 4,000 deg. C inferno as it glowed white hot. The men, who had left their spacesuits behind on Aquarius, were not let down by their craft when they were most dependent on it.

But the flight controllers were again becoming concerned. Three minutes passed, then another 30 seconds and still no word. Kerwin gave them a call, but there was no reply. Everyone began to despair. Then came the sound of Jack Swigert's voice:

"OK, Joe."

In full view of the cameras on board the carrier *Iwo Jima* first the drogues and then the three main parachutes opened. It was one of the most accurate splashdowns of all time.

Challenger and Chernobyl

The explosion of the spaceship *Challenger* and the wreckage of the nuclear power station at Chernobyl, happening in the same year, were perhaps the two most famous technological disasters in the twentieth century. Yet two experts, Dr Hans Mark and Professor Larry Carver, argue here that they will have unexpectedly hopeful consequences.*

> *You mock the very skill that proves me great*
>
> — Sophocles, *Oedipus Rex*

The first half of 1986 was not good for new technology. Two accidents, the loss of the Space Shuttle *Challenger* and the explosions at the Chernobyl nuclear power station in Russia, have raised questions about our ability to manage high technology. Though vastly different technologies were involved in accidents in two different nations with very different attitudes towards management and public information, the two events shared one important aspect: in each case a large-scale system of great technical complexity failed catastrophically.

Such accidents call attention once again to forgotten truths. Our very humanity depends on accepting the risks of failure and choosing to go on. To renew our faith in new seas and sail upon them regardless of risk is the most important lesson of these disasters. These are not easy lessons; deep-seated fears tell us to ignore them. But to do so would be to sacrifice not only our technological prowess but also the values upon which our very humanity depends.

Let us first narrate what happened and why.

Destruction came to *Challenger* 73 seconds after it was launched. Its cause was a failure in the seal of the joint between the lowest segment, and the

---

* Dr Mark's views are particularly interesting since until two years before the Challenger accident he was Deputy Administrator of NASA.

next to the lowest, in the O-rings in one of the vehicle's Solid Rocket Motors [SRMs]. But as with most technological failures, the main cause lies deeper. Many factors led to the accident. Four years before, engineers had raised questions about the O-rings. At the time, NASA management decided that these were not serious enough to warrant changes. The seed for the *Challenger* tragedy had been sown.

During the design of the Space Shuttle, NASA tried to make as many of the subsystems as possible fail-safe, that is, to design them in such a way that a single failure would not have catastrophic consequences. In this case, this was accomplished by putting two O-rings in the joint on the theory that if the first one failed, the second would do the job.

On the Shuttle's 10th flight in 1984, some charring of the O-rings was noticed. This phenomenon had been observed once before, on the second flight, but when it did not reappear, it was thought to be a one-time event. When it was seen again on the tenth flight, the problem was discussed at the Flight Readiness Review for the eleventh flight. Mark issued an "Action Item" asking for a complete review of all the SRM seals and joints. Unfortunately, this review was never held. Mark left NASA about two weeks after the signing of the Action Item, and the matter was apparently dropped.

Later events are explained in the Rogers Commission Report. The people at Marshall Space Flight Centre at Huntsville, Alabama, and the contractor Thiokol decided that they would themselves fix the O-ring problem rather than discuss it with the higher level NASA management. Consequently, nothing was done for fifteen months to make senior NASA management more aware of the problems of the O-ring seals.

Whether an investigation would have changed the decision to fly the day of the accident one cannot know. Of course, not a single flight of the Space Shuttle ever went unopposed by some group of engineers responsible for one or another of the subsystems. Sometimes NASA administrators took their advice and postponed the launch; at others they went ahead and flew anyway. Simply because a group of engineers opposed the launch from fear that a subsystem might fail was not enough to cancel it.

But photographs of the ice on the launch pad make it surprising that NASA management had given the go-ahead to fly. The launch pad structures and the gantry were completely covered in ice, and there were a great many icicles. Those icicles become missiles when the pad vibrates during take-off, and they can easily damage the heat protection tiles. The tubes that carry the liquid hydrogen for the cooling of the Shuttle's main engine nozzles are also vulnerable to flying ice. A rupture of these tubes could lead to catastrophe. NASA had often cancelled launches with much less ice on the pad than shown in those pictures.

And there clearly was a failure of communication in the decision to launch. The necessary technical information apparently never reached top-level NASA management. The problem of the O-ring failures was never properly explained to the Administrator and his assistants because people apparently thought these things were being fixed.

In the case of the decision to launch *Challenger*, the three senior people in NASA who bore the responsibility did not know that the Thiokol engineers had objected to flying because of the cold weather. Good communications between top management and engineers simply did not exist.

It would be easy to conclude that a better information system would have solved the problem. Just ensure that the engineers report everything they are doing, and all will be well. But it is not that simple. That would merely guarantee that more would be written than read! No, to correct these problems, people at the top must ask the right questions. The people in the ranks must tell the truth and write clear reports. In this, both the engineers and the NASA leadership failed on that fateful day and during the preceding months.

The *Challenger* tragedy was played out in the full glare of media attention. The whole subsequent investigation of the accident by the Rogers Commission has been published. A thorough effort was made to ensure that everything that could be known about the accident was made public.

On 26 April 1986, people near the Chernobyl nuclear power plant about 80 miles north of Kiev in the Ukraine reported fires and two explosions. We can now put together a good, if incomplete picture of what happened and why.

Unit 4 at Chernobyl, a graphite-moderated uranium reactor of 3,200 megawatts of thermal energy, produced about 1,000 megawatts of electrical energy while running. Graphite-moderated reactors are inherently more dangerous than the water-moderated ones commonly used in the United States. Graphite plants have a reactivity that increases slowly with the temperature. The reactor is thus inherently unstable and requires a complex and sensitive control system to manage its power level.

At the time of the accident, the staff were conducting an experiment. Should the reactor fail, they wanted to know how much energy the turbine rotor would contribute to auxiliary power supplies before standby diesel generators could be started. This test meant reducing the reactor's power over a 5-minute period from 100 per cent to 50 per cent, from 3,200 to 1,600 megawatts. One of the two turbogenerators was then shut down.

The plan called for a further reduction in the reactor's power to between 700 and 1,000 megawatts, but the operators apparently failed to activate on time the automatic controls that would have stabilised the reactor when operating to below 30 megawatts.

They came dangerously close to losing control of the reactor at this point. In order to increase the reactor's power, they pulled many of the manual control rods; this led to increased reactivity, and the power rose to 200 megawatts. But they were dealing with an increasingly unstable reactor core, and the prudent course would have been to halt preparations for the test and shut down the reactor.

Instead they plunged ahead. They may have felt: "Let's get this test done, procedures be hanged." Despite their cavalier treatment of the reactor, they soon brought it to a fairly stable condition. But they then committed two crucial errors. To run one test and repeat it quickly, they shut off the emergency signal system from the turbine stop valve, and they reduced the coolant to the lowest level required for the experiment.

Reactivity increased, and although a computer printout warned them of the need to shut down the reactor, they went ahead. Two things, happening simultaneously, were rushing the reactor to disaster: reactivity was increasing, and the reduced coolant was quickly letting the cooling water approach the point where it becomes steam. Control rods would now have had to go well into the core to provide the neutron absorption necessary to stabilise the reactor. It was too late.

It is probably at this point that one of the fuel elements in the reactor ruptured. The Chernobyl reactor is fuelled by slightly enriched uranium oxide pellets enclosed in zirconium-alloy fuel rods. Above 1,000 deg.C, steam reacts with zirconium to make hydrogen. Once this process started in the reactor core, there was a source of hydrogen, and the graphite moderator probably also started burning slowly, as it will in hot steam, producing carbon monoxide. Carbon monoxide can react explosively with hydrogen. The zirconium-steam reaction spreads rapidly once it starts. A source of hydrogen was thus created, with the slower graphite-steam reaction producing the other ingredient, carbon monoxide, necessary to have an explosion, or two, explosions.

The explosions seem to have completely destroyed the building of Unit 4. They also probably destroyed the remaining cooling system, and so the core of the reactor melted. The Russians deny that a meltdown occurred, but the evidence suggests otherwise. We know, for example, that temperatures in excess of 2,500 deg.C, a white-hot meltdown temperature, must have existed because isotopes of very refractory materials such as ruthenium-103 and ruthenium-106 were picked up outside the Soviet Union.

The environmental effects of the accident became apparent a few days later when a radioactive cloud drifted over Sweden. The Swedes saw radiation 100 times above the normal level, and analysed the isotopes that made it. About 40 per cent of the inventory of volatile radioactive material in the reactor core must have been released. About 80 million curies of iodine-131, the most volatile of the radioactive fission products, were dumped into the atmosphere. (A curie is roughly the radioactivity of one gram of radium.)

It was the biggest release of radioactivity ever from a nuclear reactor. The Three Mile Island accident in 1978, by comparison, caused the release of approximately 1 curie of iodine-131. The Chernobyl release was 80 million times larger, and at least one-third of that from Hiroshima.

The Chernobyl accident has so far killed 31 people. These were staff and firemen, dead from the shock effect of the explosion, by secondary effects like collapsing buildings, and by very high radioactivity. Another 170 or so have been hospitalised with acute radiation sickness, and many will perish before their time.

About 100 million people in the western Soviet Union and Eastern Europe were exposed to higher than normal radiation. In the next 50 years, there will be about 10,000 extra cancer cases among them.

In addition to these direct effects, about 14,000 people have been evacuated from the Chernobyl area. The exclusion zone around Unit 4 will have to be maintained for years. The Russians have stopped people from living in a region about 15 miles around the destroyed reactor.

The immediate cause of the Chernobyl catastrophe was human error compounded by the pressure to finish a task quickly. Mr Valery Legasov, a Soviet spokesman, has admitted that the operations could override safety precautions. But whether from ignorance of the basic physics of the reactor, out of pressure to do the job, or both, they shut down automatic safety systems to conduct an experiment without proper authorisation. That the reactor lacked multiple shutdown systems allowed human errors to lead to disaster.

In contrast with the *Challenger* accident, where the whole drama was played out in full public view, the Russians were acutely embarrassed by the piecemeal way in which the events at Chernobyl came to public attention. They had no tradition of operating openly, as did the nuclear industries in the West. They improvised. Their first instinct was probably to keep as much as possible secret. That they were finally unable to do this, and that they have been as forthcoming as they have, is perhaps a hopeful sign. For if high technology involves risks in an open society, those risks are increased in a closed one. This is but one of the lessons of *Challenger* and Chernobyl.

Challenger and Chernobyl confirm Dr Johnson's observation that men more often need to be reminded than taught. Both disasters hold out vital lessons.

The first is that ignorance is the father of disaster, a father whose progeny multiply hideously in powerful and complex technologies. Ignorance of two kinds contributed to these tragedies. First, the engineers in NASA did not really understand how the lower field joint on the SRM actually worked. Equally clearly, the Chernobyl engineers did not understand the control system of their reactor. In both cases, this ignorance was compounded by flawed decision-making. It is not enough to know; we must communicate what we know. Ignorance about how people and machines interact as well as about how people interact with people led to both catastrophes.

The second lesson is that discipline is vital. This means the obligation of engineers to report honestly and of the management to ask the right questions. In the case of the decision to launch *Challenger*, the necessary discipline broke down. It is likely that the same breakdown in discipline occurred at Chernobyl, that breakdown perhaps being much worse. There is some evidence, for example, that the leadership in Moscow did not know what was happening until radioactive debris was detected outside the Soviet Union. They were probably told of an accident, but the seriousness of the situation was withheld from them. That they waited for three days to carry out the massive evacuation around the area suggests this and points to a very great failure in discipline.

That we need information and the discipline to use it are hardly startling truths, and why we need disasters to demonstrate them remains a mystery. Our guess is that when the space shuttles fly again, their flights will be better and safer than in the past.

This conclusion is in accordance with history. The Apollo fire in 1967 caused a similar situation. There was a delay in space flights. But after a thorough investigation, both the technology and its management improved. NASA met the objective set by President Kennedy to put a man on the Moon by 1970 and to bring him back safely. We believe that NASA will build the Space Station as planned, and the benefits promised by space exploration will become tomorrow's realities.

The lessons from Chernobyl could lead to equal optimism for nuclear power. Though this may seem paradoxical, the Chernobyl accident, we believe, will make it easier to build nuclear power reactors, partly because the catastrophe at Chernobyl was the "maximum credible accident" that the U.S. Nuclear Regulatory Commission requires people who operate reactors to analyse. This was the first such accident. The one at Three Mile Island had been the only existing precedent. The opponents of nuclear power thus had the industry over a barrel. They could invent any and all terrible consequences of a "maximum credible accident" that they chose.

This will not be true from now on. The Chernobyl disaster is, without question, a "maximum credible accident." Thus we now have a data base to make better judgments. What will its consequences be when all the facts are known? About 200 people will die of exposure to radiation. The immediate effect of the accident is thus similar to a major airliner crash. If this situation is confronted openly, the public perception of nuclear reactor problems will surely become more accurate.

With these results of *Challenger* and Chernobyl, we will reap further insights in the exploration of space and of nuclear energy. And they will affirm the values of openness, of confronting failure, and of taking responsibility for it. The consequence of *Challenger* — in the United States at least — has been catharsis, a healing and a renewal of strength.

The relative openness of the Soviet Union towards the Chernobyl accident, as shown by their report to the International Atomic Energy Agency in Vienna, has been a positive relief. The Russians may have recognised that the price for not being open is a neurosis that not only cripples science and engineering but also absorbs creative energies. The values of openness, responsibility, and rationality are inimical to a totalitarian society. Its tendency is to treat error, not as an opportunity to learn, but as a sign of weakness, something to be hidden rather than as something to be exploited for future strength. We can only hope that the investigation into Chernobyl will become the rule, not the exception, in Russian life and that they go on to confirm the intuition of James Joyce that people of genius make no mistakes. "Their errors are volitional and are the portals to discovery."

It is a fearful thing to confront the unleashed powers of the Universe. But the danger is that nuclear power, its risks and rewards, will not be weighed rationally; instead, that irrational fears will exile it to some *terra incognita* of the mind in the hope of forgetting it: that the fate of nuclear power will not be decided by the best information available but by dark passions.

This possibility should come as no surprise. Most people never think clearly about technology. Their thoughts are usually intermingled with fears of death and a sense that we do not know where we are going. Both of these emotions are, as the protests against nuclear power show, very strong.

The same argument applies to space. If we should stop our ventures into space out of fear of risk — a fear really of freedom and of death — then we will have sacrificed more than the space programme. We will have given up what makes us human: the capacity to confront and weigh our fears and to act despite them.

To face a tragedy openly; to investigate it regardless of the consequences; to conclude that though man can overcome some of the problems, failure will inevitably come again; to confront all of this, painful as it has been, and to go on — all this testifies to an open, rational and courageous spirit.

But as *Challenger* reveals our openness to light, the reaction to Chernobyl may reveal our susceptibility to the forces of darkness. While working on this paper, we asked several people about their reaction to nuclear power. One simply said: "I hope you don't begin to glow." It was a humorous manifestation of a deep irrational response to nuclear power, like the death masks that so often accompany anti-nuclear demonstrations. Such fears need to be understood and defused.

If we overcome present fears, if we go on to exploit nuclear power for the betterment of man's life, it may well be because the Russians, uncharacteristically, have been open about Chernobyl. Attempts to remain secretive would have led, and did for a time, to greater fear and a dimmer future for nuclear energy.

In *Challenger* and Chernobyl, we have rehearsed again the tragedy of Sophocles. The people of Thebes wanted to stop the plague that was destroying their city. Some were ignorant of the cause; others wanted to keep it hidden. Oedipus alone could not "leave the truth unknown." He knew that hiding the truth would prolong the pestilence and destroy the state. Seeking the truth was dangerous, but the search for it was his greatness. Searching for truth and becoming human are inextricably tied together. Oedipus's quest, in which his greatest obstacle is the fear and ignorance of others, leads to great pain and loss; the very thing that makes him great leaves him vulnerable. He finds himself, as we do with *Challenger* and Chernobyl, "mocked for the very skills that prove me great." He could have ignored the truth and given in to fear. We can too. But to do so is not to be human.

NOTES

The principal source for the *Challenger* accident and its causes was the report of the Rogers Commission which investigated it. See *Report of the Presidential Commission on the Space Shuttle Challenger Accident* (Washington DC, 6 June 1986.)

The main source for Chernobyl is the highly detailed, though somewhat speculative, *The Accident at the Chernobyl Nuclear Power Plant and its Consequences, Information compiled for the International Atomic Energy Authority Experts Meeting in Vienna, 25–29 August, 1986.*

3
The
Deep Skies

The World Moves!

The Polish astronomer Nicholas Copernicus wrought the most fundamental revolution in science of all times — and he even invented the word 'revolution.'

It occurred to him that tables of planetary motion would be rendered more intelligible if one assumed that the Sun, rather than the Earth, was the centre of our solar system, and that all the planets, including the Earth, revolved around it. He then proved, in his 1543 book *On the Revolutions of the Earth through the Heavens*, that this was actually the case.*

His proof was exceedingly simple. The backward or 'retrograde' motions of Mars, Jupiter and Saturn had baffled all previous astronomers because they moved in the opposite direction to that of the stars. Copernicus realized that this could be explained because they were farther from the Sun than was the Earth, which would therefore continuously overtake them in its orbit.

No one ever dealt the false pride of mankind a heavier blow. Realizing that the Church would bitterly resent his theory, he carefully dedicated his book to the Pope:

To His Holiness Paul III, Supreme Pontiff.

---

* The idea was not entirely new. Aristarchus of Samos had tentatively suggested it seventeen centuries before, as had Nicholas of Cusa, in Copernicus's own time. But to Copernicus belongs the fame of working it out in detail.

I can certainly well believe, most holy Father, that while perhaps a few will accept this book of mine about the revolutions of the Earth, some will clamour that I ought to be cast out for writing it. Nevertheless, I think that opinions wholly alien to what is right ought themselves to be driven out. Yet when I reflected on the absurd fairy tale that people have been brought up on through the ages, that the Earth is motionless in the midst of the heavens, as if it were the centre of it, I hesitated long before asserting that on the contrary it moved.

Indeed, I sometimes wondered whether it would be better to imitate the Pythagoreans and others who passed down knowledge, not by books but from hand to hand, and only to their friends and relatives. But they seem to have done this lest their discoveries should be despised by those who think writing a book is pointless unless it makes a profit. And there were those who were enthusiastic about imparting knowledge but were too stupid to be any more usefully engaged on it than drones among bees. When therefore I had pondered these matters, the scorn to be feared from the novelty and the supposed absurdity of my opinions impelled me to set aside all work on the book.

But friends persuaded me otherwise. Among the foremost of these was Nicholas Schonberg, Cardinal of Capua, celebrated in all fields of learning. Next was Tiedemann Giese, Bishop of Kulm, most learned in all sacred matters. He has repeatedly urged me, and sometimes even with censure, to publish this book that I have held back for more than nine years. Other eminent men also pleaded with me, urging me no longer to refuse to contribute to public knowledge from an imagined dread. They said that however absurd my doctrine of the Earth's motion might at first appear, so much the greater would be the admiration and the goodwill after people had seen the mists of absurdities rolled away. Encouraged by these pleaders, I at last permitted them to publish my work.

Your Holiness may particularly wish to hear how, contrary to accepted mathematics and to common sense, I dared to imagine that the Earth moved. Yet mathematicians disagreed among themselves about the reasons for the movements of the planets. They are even so completely undecided about the motion of the Sun and the Moon that they could not prove the constant length of the great year.* When questioned about these matters, they showed that they did not know what they were talking about. For those who believed in planets circling a common centre at the Earth could find no phenomena that such circles would explain. But those who maintained that all the circles must be eccentric have had to admit that there are enough exceptions to make their assertion nonsensical.

Moreover there is a definite symmetry in the universe which they could find no evidence to explain. In their 'systems', it is as if a man should acquire his hands, his feet, his head and his limbs, all from different bodies. The resulting creature would resemble a monster. Thus, in each of their systems, they are found to have passed over some essential, or to have admitted something both strange and irrelevant. This would not have happened if they had followed definite principles.

Having long brooded on the chaos of traditional mathematics, it began to weary me that no more definite explanation of the movements of the world machine existed among the philosophers who had studied so exactly *in other respects* the minutest details of the world. Wherefore I re-read the books of all the philosophers that I could obtain, to find out whether anyone had ever conjectured that the motions of the spheres might be not as taught in the schools. And I first found that, according to Cicero, Nicetas [or Hicetas] had thought the Earth moved. Then I found in Plutarch that others had held the same opinion:

> But while some say the Earth stands still, Philolaus the Pythagorean held that it moves about the central fire in an oblique circle, after the same manner of motion that the Sun and Moon have. **Heraclitus of Pontus and Ecphantus the Pythagorean assign a motion to the Earth, not progressive, but after the manner of a wheel being carried on its own axis. Thus the Earth, they say, turns upon its own centre from west to east.

I myself then began to meditate upon the mobility of the Earth. And although the opinion seemed absurd, yet because I knew of the liberty

---

* The 'solar year', first discovered by Hipparchus in about 100 BC. It is the 26,000 or so Earth-years in which the stars complete their apparent circuits round the Earth and return to their original positions. This phenomenon, known as the precession of the equinoxes, occurs not because of any movement of the stars but because of gradual change in the direction of the Earth's axis of rotation.

** The 'central fire' here is not the Sun, but the centre of the universe, around which, according to the Pythagoreans, all celestial bodies rotated.

granted to others to imagine whatsoever circles they pleased to explain the phenomena of the stars, I thought that I also might readily be allowed to seek proofs of the Earth's motion.

Thus supposing these motions which I attribute to the Earth, I found at length by long observation, that if the motions of the other planets were added to the rotation of the Earth, this would explain their order of appearances so clearly that *no part of model could be altered without creating confusion in the rest of it.*

The motions and appearances of the planets, hitherto mysterious, become clear if we attribute motion to the Earth. I cannot doubt that skilled mathematicians will agree with me if they will but examine and judge, not casually but deeply, the evidence I have gathered to prove these things. So that learned and unlearned may alike see that in no way do I evade judgement, I prefer to dedicate these my lucubrations to your Holiness rather than to anyone else; especially because even in this very remote corner of the Earth in which I live, you are held so very eminent by reason of the dignity of your position and also for your love of all letters and of mathematics that, by your authority and your decision, you can easily repress the malicious attacks of calumniators.*

If perchance there should be foolish speakers who, together with those ignorant of all mathematics, will take it on themselves to decide concerning these things, and because of some place in the Scriptures wickedly distorted to their purpose,** should dare to assail this my work, they are of no importance to me. I despise their judgements as rash. I would compare them with Lactantius, the celebrated writer who proclaimed the Earth was flat and spoke so childishly of those who said it was spherical. So it will not be surprising if such should laugh at me also.

Mathematics may be written for mathematicians, but if I am not mistaken it will contribute something even to your Holiness's state. For it is not so long ago under Leo X when the question arose in the Lateran Council about correcting the Ecclesiastical Calendar. It was left unsettled then for one reason — that the length of the year and of the months, and the movements of the Sun and Moon, had not been satisfactorily determined. From that time on, I have turned my attention to the more accurate observation of these, at the suggestion of that most celebrated scholar, Father Paul, a bishop from Rome, who was the leader then in that matter. What, however, I may have achieved in this, I leave to the decision of your Holiness especially, and to all other learned mathematicians.

Nicholas Copernicus

---

* There is no evidence that Paul III ever reacted to Copernicus's book or its dedication. He was preoccupied with Papal and European politics, and showed no interest in science.
** Copernicus presumably refers to the story of Joshua halting the Sun in its course. It was this passage in his letter that so infuriated Martin Luther, who declared his theory to be 'anti-Biblical and intolerable'.

The Story
of Tycho Brahe

The Danish nobleman Tycho Brahe was probably the finest astronomical observer before the age of telescopes, as Patrick Moore
explains:

Tycho Brahe, born in 1546, three years after the death of Copernicus, had
the same love of astronomy as that great theorist, and the same urge to find
out as much as he could.

He was of noble ancestry; his father Otto was Governor of Helingbourg
Castle, and a man of great influence. Early events set the pattern for his
tempestuous life. For some unknown reason Otto had promised his
brother Jorgen, an officer in the Danish Navy, that as soon as he had
another son he would hand over little Tycho to be brought up in Jorgen's
household. When another boy was born, Otto had second thoughts, with
the result that Uncle Jorgen kidnapped Tycho.

The boy's upbringing at Jorgen's home was probably conventional
enough. In 1559 he was sent off to Copenhagen University, where he saw a
partial eclipse of the Sun which may have started his love of astronomy.

Star-gazing did not suit his uncle, who was set upon making him an
important figure in Danish politics. The law was the chosen study, and
Leipzig University had an excellent law school. But Jorgen did not trust
Tycho's dedication to legal studies, and probably cast unfavourable looks
at the copy of Ptolemy's *Almagest* which the boy had bought when still at
Copenhagen. So a companion, Anders Vedel, was sent with him to Leipzig,
to keep his mind upon the law. It was an impossible task, and Vedel had to
agree to a compromise; the student could look at the stars by night
provided he put in a full day's work.

In 1563 there was a conjunction of the two bright planets Jupiter and
Saturn; that is to say, they appeared side by side in the sky. Tycho realized
that the date of the conjunction given in the old planetary tables of King
Alphonso X of Castile* was a full month wrong, and that even the more
accurate tables of Erasmus Reinhold compiled 12 years earlier were in
error by several days. This was his first recorded observation, and he made
it in a decidedly primitive way. His only equipment consisted of a pair of

---

* Alphonso (1221–84) is famous also for the remark 'Had I been present at the Creation, I
would have given some useful hints for the better ordering of things.'

compasses. By holding the centre close to his eye, and pointing one arm to Jupiter and the other to Saturn, he was able to find the angular distance between them. This did not satisfy him, and he began to make proper measuring instruments.

Jorgen died in 1565 from a curious accident; he caught pneumonia after soaking himself in pulling the King of Denmark out of a moat. Tycho promptly abandoned his legal studies. He went to Rostock University in Germany, where he fought a duel with another student, in which, in the best medieval tradition, part of his nose was sliced off. He repaired the damage with gold, silver and wax, and this was how his nose remained for the rest of his life, apparently without causing him any discomfort.

The year 1571, when Tycho was 25, marked the turning point of his life: the appearance of a brilliant new star in the constellation of Cassiopeia. It blazed out in November. He described it in his book *De Stella Novis* (On the New Star):

> "In the evening, after sunset, I noticed a new and unusual star, surpassing all the others in brilliancy, shining almost directly above my head. Since I had, from boyhood, known all the stars in the heavens perfectly, it was quite evident to me that there had never before been a star in that place, even the smallest, to say nothing of a star so conspicuously bright as this. But when I observed that others, too, could see it, I had no further doubts. A miracle indeed, either the greatest that has occurred in the whole range of nature since the beginning of the world, or one certainly that is to be classed with those attested by the Holy Oracles."

What made this star so important? Simply the fact that according to Aristotle, the heavens and the starry sphere were changeless. Yet this new star was more than normally intrusive; it could be seen in daylight! We now know it as Tycho's Star.

It was nothing more nor less than a supernova, not as brilliant as that in the constellation of Lupus in 1006, but fully equal to the 1054 outburst in Taurus. Tycho had no idea of its real nature, but he made precise measurements of it. He found no detectable change in its position, and he concluded that it must be very remote. As the weeks went by, it faded. By December, 1572, it was about equal in brightness to Jupiter; by the following March it was still fading, and a year later it disappeared. During the decline, its colour changed, and Tycho, as superstitious as any astronomer of his time, was not slow to point out the astrological implications:

> 'The star was first like Venus and Jupiter, giving pleasing effects; but as it then became like Mars, there will next come a period of wars, seditions, captivity and deaths of princes and destruction of cities,

together with dryness and fiery meteors in the air, pestilence and venomous snakes. Lastly, the star became like Saturn, and there will finally come a time of want, death, imprisonment and all sorts of sad things.'

By now Tycho was becoming well-known, and in 1576 King Frederick II of Denmark offered him a full-scale observatory and funds to maintain it. The chosen site was Hven (now called Ven), a low-lying island in the Baltic. Tycho accepted, and in 1576 he began the construction there of his observatory — Uranibourg, the 'Castle of the Heavens.' It was built in the middle of a large square enclosure laid out as a garden whose corners pointed north, south, east and west. It contained a library and a chemical laboratory as well as living quarters and the rooms for the instruments themselves. Later, in 1584, he built a second 'Castle of the Stars' on the island, Stjerneborg, in which some of the instruments were located below ground level, because, when the wind blew strongly they were shaken to and fro.

Tycho lived in magnificent style. Guests from all over Europe, including the future James I of England, were royally entertained with banquets, games and hunts. But the islanders were less well treated. Tycho was a harsh landlord, and even built a prison for those tenants who failed to pay their rents.

His instruments were by far the best of their time, and being a most careful and accurate observer he obtained excellent results. He measured the positions of 777 stars, and drew up a catalogue; his star positions were never in error by more than one or two minutes of arc. When we remember that he had no telescopes, we can see how good an astronomer he must have been. Yet he was always devoted to astrology, and he never began observing without dressing himself in special robes.

He measured the movements of the planets, and these observations later enabled his student Johannes Kepler to construct his law of elliptical planetary orbits. Tycho observed seven comets during his stay at Hven, correctly stating that they were much more distant than the Moon, thereby disposing of Aristotle's theory that comets were 'atmospheric exhalations' a few miles up. But Tycho could never accept Copernicus's theory of a moving Earth, partly on religious grounds but mainly because he believed that on the Copernican system the stars would have to be tremendously remote — which we now know they are. He preferred a kind of hybrid system, in which the planets revolved around the Sun, while the Sun and Moon were in orbit round the Earth.

Unkind people have referred to this system as Tycho's Folly, but the great Dane was inordinately proud of it, and reacted furiously when he believed that it had been copied and re-presented by an avowed enemy, Reymers Bar, a mathematician who had begun life as a swineherd.

More mundane matters had begun to intrude on his life at Hven, and his fruitful period was coming to an end. King Frederick died in 1588, and soon Tycho lost another ally, the Danish Chancellor Niels Kaas. His hot temper, his ill-treatment of his tenants, and his neglect of official duties led to a break with the court. His funds were cut off.

Tycho abandoned Hven, taking the main observing instruments with him, and eventually arrived in Prague, the capital of Bohemia, to become Imperial Mathematician to the Holy Roman Emperor Rudolph II. The castle of Benatek, outside Prague, was placed at his disposal, but conditions there were very different from those at Hven. The Emperor was mainly interested in alchemy (the so-called making of gold out of baser elements). His reign was a succession of disasters, and finally, in 1611, he was deposed. Before long Betanek was given up through lack of money, and Tycho settled in Prague itself. On November 24, 1601, he died suddenly, at the age of 55.

Tycho Brahe was hasty, intolerant, unashamedly egotistic and often cruel. But he was at the same time brilliantly clever, sincere and hard-working.

Nothing now remains of Uranibourg and Stjerneborg. The site of Tycho's castle at Hven is occupied by a grassy dip, with only a few markers to indicate where the building once stood. Yet when I visited Hven a few years ago, I found Tycho's statue, vast, commanding and powerful, surveying the scene where so much pioneering work had been done. The spirit of the Master of Hven lives on.

The Misdeeds of Isaac Newton

Isaac Newton was a colossus. He discovered all the laws which make spacecraft and satellites possible, and that which prevents the Moon from falling on our heads. But there was a dark side to him, as Stephen Hawking relates:

Newton was not a pleasant man. His relations with other academics were notorious, with most of his later life spent embroiled in heated disputes. Following the publication of his *Principia Mathematica* — surely the most influential book ever written in physics — Newton had risen rapidly into public prominence. He was appointed president of the Royal Society and became the first scientist ever to be knighted.

Newton soon clashed with the Astronomer Royal, John Flamsteed, who had earlier provided Newton with much needed data for his *Principia*, but was now withholding information that Newton wanted. Newton would not take no for an answer; he had himself appointed to the governing body of the Royal Observatory and then tried to force immediate publication of the data. Eventually he arranged for Flamsteed's work to be seized and prepared for publication by Flamsteed's mortal enemy, Edmund Halley. But Flamsteed took the case to court and, in the nick of time, won a court order preventing distribution of the stolen work. Newton was incensed and sought his revenge by systematically deleting all references to Flamsteed in later editions of the *Principia*.

A more serious dispute arose with the German philosopher Gottfried Leibniz. Both Leibniz and Newton had independently developed a branch of mathematics called calculus, which underlies most of modern physics. Although we now know that Newton discovered calculus years before Leibniz, he published his work much later. A major row ensued over who had been first, with scientists vigorously defending both contenders. It is remarkable, however, that most of the articles appearing in defense of Newton were originally written by his own hand — and only published in the name of friends! As the row grew, Leibniz made the mistake of appealing to the Royal Society to resolve the dispute. Newton, as president, appointed an 'impartial' committee to investigate, coincidentally consisting entirely of Newton's friends! But that was not all: Newton then wrote the committee's report himself and had the Royal Society publish it, officially accusing Leibniz of plagiarism. Still unsatisfied, he then wrote an anonymous review of the report in the Royal Society's own periodical. Following the death of Leibniz, Newton is reported to have declared that he had taken great satisfaction in 'breaking Leibniz's heart.'

During the period of these two disputes, Newton had already left Cambridge and academe. He had been active in anti-Catholic politics at Cambridge, and later in Parliament and was rewarded eventually with the lucrative post of Warden of the Royal Mint. Here he used his talents for deviousness and vitriol in a more socially acceptable way, successfully conducting a major campaign against counterfeiting, sending several men to their deaths on the gallows.

Nothing for Nothing

One of the longest-lasting of all legends has been that of perpetual motion. The idea of building a Perpetual Motion Machine, an engine that will run for ever without energy being put into it, surpasses in persistence even the myth that chemistry can turn iron into gold.*

But perpetual motion is eternally impossible because it violates the Second Law of Thermodynamics, discovered independently by Lord Kelvin and Rudolf Clausius, which predicts that every single thing that exists suffers an inexorable loss of heat, and the corresponding increase of chaos, or 'entropy'. Since heat is equivalent to energy, it follows that any conceivable machine, no matter how efficiently designed, will run down when it loses power. Sir Arthur Eddington had this to say about the Second Law:

The law that entropy increases — the Second Law of Thermodynamics — holds, I think, the supreme position among the laws of Nature. If someone points out to you that your pet theory of the Universe is in disagreement with Maxwell's equations — then so much the worse for Maxwell's equations. If it is found to be contradicted by observation — well, these experimentalists do bungle things sometimes. But if your theory is found to be against the Second Law of Thermodynamics I can give you no hope; there is nothing for it but to collapse in deepest humiliation.

'The Learned Astronomer'

When I heard the Learn'd Astronomer,
When the proof, the figures, were ranged in columns before me,
When I was shown the charts and diagrams, to add, divide and
  measure them,
When I sitting heard the astronomer where he lectured with
  much applause in the lecture room,
How uncomfortable I became, tired and sick.
Till rising and gliding out I wander'd off by myself
In the mystical moist night air, and from time to time
Looked up in perfect silence at the stars.

— Walt Whitman, 'The Learned Astronomer'

---

* In the cores of massive stars, this ceases to be a myth. At temperatures of hundreds of millions of degrees, lighter metals *will* fuse into gold. That is why gold exists in the Universe.

Einstein and Time Travel

In 1887 two American physicists, Albert Michelson and Edward Morley, performed an experiment which had totally negative results, but which produced the most shattering consequences.

It was ultimately to reveal why it is impossible to exceed the speed of light, and of the strange things that would happen to astronauts who even approached that speed.

Michelson and Morley, like all physicists of the time, assumed that space was filled with a mysterious and invisible substance called 'luminiferous ether', meaning that it was supposed to carry light waves. Otherwise, they reasoned, how could light-waves travel?

In a basement in Cleveland, Ohio, they set up an experiment with light beams to measure the speed of Earth's passage through this stationary ether. Because of the Earth's movement through space, light-waves would surely travel slightly more slowly *in the direction of the Earth's motion* through the ether than they would when going away from it – just as a pilot in an open cockpit would feel the wind against his face.

To their amazement and everyone else's, they found that light always travelled at exactly the same speed, 186,282 miles per second (299,792 kilometres per second), irrespective of the Earth's motion and *irrespective of the speed of its source.**

Physicists found it hard to believe that if there was an ether, nature would go to such drastic, almost prankish, lengths to conceal its existence. Bertrand Russell likened the behaviour of nature in this case to that of the White Knight in *Alice Through the Looking Glass:*

> *But I was thinking of a plan*
> > *To dye one's whiskers green,*
> *And always use so large a fan*
> > *That they could not be seen.*

Not only was there no ether; there was another mystery as well. What consequences follow if light behaves in this inexplicable way? Albert Einstein answered this question in 1905 with his Special Theory of Relativity. If light *always* travels at the same speed, he

---

* The Michelson–Morley experiment was no freak. It has been repeated many times by others, and always with the same results.

said, irrespective of whether you are travelling towards its source or away from it, then at very high speeds your clocks must slow down. This so-called Twin Paradox is perhaps the most interesting aspect of his Special Theory. Martin Gardener explains:

The Twin Paradox is a thought experiment involving two twins. They synchronize their watches. One stays at home while the other boards a spaceship and makes a long journey. When the space traveller returns, they compare watches. According to the Special Theory of Relativity, the traveller's watch will show a slightly earlier time. In other words, time on the spaceship will have passed more slowly than time on Earth.

So long as the journey is confined to the solar system, and made at relatively low speeds, this time difference will be negligible. But over long distances, with speeds close to that of light, this so-called 'time dilation' can be large. Someday a means may be found to accelerate a spaceship until it reaches a speed only a trifle below that of light. This would make possible trips to other stars in the Galaxy, and perhaps even to other galaxies. So the Twin Paradox is more than just a parlour puzzle; it may one day become a common experience of space travellers.

Suppose that the astronaut twin goes a distance of 1,000 light-years* and returns: a small distance compared with the diameter of our Galaxy. Would he not surely die of old age long before he completes the trip? Would not his trip require, as in so many science fiction stories, an entire colony of people so that generations would live and die while the ship was making its long interstellar voyage?

The answer depends on how fast the ship goes. If it travels at just under the limiting speed of light, time within the ship will proceed at a much slower pace. Judged by *Earth-time,* the trip will of course take more than 1,000 years. But judged by the astronaut on the ship, if he travels fast enough, it may take only a few decades!

Here is a simple calculation: An astronaut travels out from Earth to the great spiral galaxy in Andromeda, 2 million light-years away, accelerating as he goes to a cruising speed of more than 99.9999... per cent of the speed of light. The ship will be going so fast that the astronaut ages only 5 years while making it. But when he returns, he will find that on Earth 4 million years have gone by!

This raises all sorts of fascinating possibilities. A scientist aged 40 and his teenage laboratory assistant fall in love. They feel that their age difference rules out marriage. So off he goes on a long space voyage, travelling at close to the speed of light. He returns, aged 41. Meanwhile on Earth, his girlfriend has become a woman of 33. Perhaps she could not wait 15 years

* A light-year is the distance that light travels in one year. It is about 6 trillion miles. Light-years are essential in astronomy. Because of the vastness of the Universe, it would be absurd to measure the distances between the stars in miles.

for her lover to return; she has married someone else. The scientist cannot bear this. Off he goes on another long trip. Moreover, he is curious to learn if a certain theory he has published will be confirmed or discarded by future generations. He returns to Earth aged 42. His former girlfriend is long since dead. What is worse, his pet theory has been demolished. Humiliated, he takes an even longer trip, returning at the age of 45 to see what the world is like a few thousand years hence.

Perhaps, like the time traveller in H.G. Wells's story *The Time Machine*, he will find that humanity has become obsolete. Wells's time machine could go both ways, either forwards into the future or backwards into the past. But this scientist can only travel into the future. Now he is stranded. He has no means of getting back into the stream of human history where he belongs.

Unusual moral questions would arise if this sort of time travel becomes possible. Is there anything wrong, for instance, in a girl marrying her own great-great-great-great grandson?

Please note: this kind of time travel avoids all the logical traps that plague science fiction, such as dropping into the past to kill your parents before you are born, or whisking into the future and shooting yourself between the eyes. Consider the limericks:

> *There was a young lady named Bright,*
> *Who travelled much faster than light.*
> *She set out one day*
> *In the relative way,*
> *And returned on the previous night.*

> *The lady was Bright but not bright,*
> *And she joined in next day in the flight;*
> *So then two made the date,*
> *And then four and then eight,*
> *And her spouse got the hell of a fright.**

By returning on the previous night, the lady must have encountered a duplicate of herself. Otherwise it would not have been truly the night before. But there could not have been two Miss Brights the night before because the time-travelling Miss Bright left with no memory of having met her duplicate yesterday. So there is a clear-cut logical contradiction. Time travel of *that* sort, which involves impossible journeys faster than light, is not logically possible unless we assume the existence of parallel worlds, running along different time tracks.

(Only a science fiction writer would dare to speculate on what astronauts might observe from a ship that moved faster than light. Perhaps

---

* The first of these two verses was written by E. Reginald Buller, of the University of Manitoba; and the second by J. H. Fremlin, of Birmingham University.

the cosmos would appear to turn inside out and become its own mirror image, stars would acquire negative mass and cosmic time would run backwards.* If the speed of light is exceeded, the equations of the Special Theory give values to length, time and mass that are what mathematicians call 'imaginary numbers' that involve impossible concepts like the square root of minus one. Who knows? Maybe a ship that broke the light barrier would plunge straight into the Land of Oz!)

The most common objection to the Twin Paradox is a refusal to accept that there is any difference in status between the space traveller and the stay-at-home on Earth. The Earth is travelling through the universe just like the spaceship. Why should one age more slowly than the other?

Einstein's answer is that the astronaut moves under his own acceleration; but when the Earth moves, *the entire universe moves with it.* Clocks on board the astronaut's ship, because it has been accelerated, are slowed down by the gravitational field of the universe; for gravity has a slowing effect on clocks. But the Earth does *not* move relative to the universe. There is therefore no gravitational field to slow down its clocks.

With two previously synchronized clocks stating different times, it cannot be stressed too often that it is incorrect to ask which one is 'right.' Both times are 'right.' There is no way to decide the question. Isaac Newton took it for granted that one universal time permeated the entire cosmos. But to Einstein there are only local times. On Earth, everyone is

---

* One physicist, Gerald Feinberg, has proposed the existence of a particle call the tachyon (based on a Greek word meaning 'swift') that *always* travels faster than light and never slows down. It would obey Einstein's equations by always travelling backward in time, from the future into the past. No tachyons have yet been detected. — A.B.

being carried along through space at the same speed, and their watches run on the same 'Earth time.' But when different speeds are involved, it is not possible to say which of two events happened before or after the other.

Afterword: The Special Theory predicts other strange effects as well. Since the speed of light in the vacuum of space is constant at 186,282 miles per second, irrespective of the speed of its source, the *length* of a spaceship in its direction of motion, just like time on board it, must therefore be reduced in proportion to its speed. And the spacecraft's *mass* — i.e., the energy required to accelerate it — is increased as its speed increases.

The extent of these changes is calculated by three simple formulae, worked out by Hendrik Lorentz and George Fitzgerald, which Einstein incorporated into his theory. To calculate the *length* of a moving object we multiply its length at rest — that is, when stationary on a planet — by

$$\sqrt{1 - \frac{v^2}{c^2}}$$

where v is the speed of the spacecraft, and c is the speed of light. Increasing *mass* of the moving spacecraft is calculated by a slightly different formula. Its mass at rest is multiplied by

$$\frac{1}{\sqrt{1 - \frac{v^2}{c^2}}}$$

Measuring the slowing of time on board the spacecraft is equally simple. To see how much more slowly the astronauts are ageing, we multiply a given period — say 60 seconds of Earth time — by the same formula as that dealing with length:

$$\sqrt{1 - \frac{v^2}{c^2}}$$

Two spaceships cannot recede from each other at a combined speed greater than that of light. If a man standing on the ground sees one craft racing overhead going due north at 90 per cent of the speed of light, and another going due south at the same speed, he might suppose that each craft was receding from the other at a combined speed of 180 per cent of that of light. Yet he would be wrong. The sum of the two speeds cannot exceed that of light. It must be calculated by the formula:

$$\frac{a + b}{1 + \dfrac{ab}{c^2}}$$

where a and b are the respective speeds of the two craft, and c is the speed of light. It will be seen from this formula that if the two craft had been travelling very slowly — at say 600 mph — then the sum of their speeds would be about 1,199.9999999 mph, or *almost* 1,200 mph. But if their speeds are very great, the formula gives a quite different kind of answer. Pretend that instead of two spaceships the man sees two light-beams receding in opposite directions. He will estimate their mutual speed of recession as twice the speed of light, or 2c. But if he were riding on one of the beams, he would estimate the other's speed of recession, according to the formula, as

$$\frac{c + c}{1 + \dfrac{c^2}{c^2}}$$

which of course works out as c.

Let us now test an imaginary spacecraft at varying speeds, and see how its length, ageing and mass change according to Einstein's equations. We will assume that a ship, while stationary in port, is exactly 100 metres in length, and has a rest-mass of 100 tons. As the ship accelerates we see that a ship-hour becomes a progressively smaller fraction of an Earth-hour.

Speed of ship as percentage of light	Length of ship (metres)	Mass of ship (tons)	Duration of ship-hour in minutes (Earth = 60)
0	100.00	100.00	60.00
10	99.50	100.50	59.52
20	97.98	102.10	58.70
30	95.39	104.83	57.20
40	91.65	109.11	55.00
50	86.60	115.47	52.10
60	80.00	125.00	48.00
70	71.41	140.03	42.75
80	60.00	166.67	36.00
90	43.59	229.42	26.18
95	31.22	320.26	18.71
99	14.11	708.88	8.53
99.9	4.47	2,236.63	2.78
99.997	0.71	14,142.20	1.17
100	zero	infinity	zero

As these figures plainly show, no spacecraft could ever travel at the speed of light itself. An infinite mass would require an engine of infinite power for its propulsion. And even if this were miraculously achieved, the spacecraft's length would be zero, and it could not therefore exist.

It is interesting that Michelson, whose experiment with Morley started the work that led to this tremendous break-through, could never bring himself to accept relativity. As he wrote in 1894:

'The more important fundamental laws of physical science have all been discovered, and these are now so firmly established that the possibility of their ever being supplanted by new discoveries is exceedingly remote. Our future discoveries must be looked for in the sixth place of decimals.'

'I'll Bring You to Your Senses!'

While it is now well established that the universe had a beginning in the Big Bang some 15,000 million years ago, this consensus was not reached without bitter controversy. Barbara Gamow, wife of the astronomer George Gamow, composed these verses about the angry dispute between Sir Martin Ryle and Sir Fred Hoyle.

Ryle's observations of objects at the edge of the cosmos convinced him of the truth of the Big Bang, but Hoyle believed (and still believes at the time of writing) in an eternal cosmos, the 'steady state' universe.

> 'Your years of toil,'
> Said Ryle to Hoyle,
>> 'Are wasted years, believe me.
> The steady state
> Is out of date
>> Unless my eyes deceive me,

'My telescope
Has dashed your hope;
   Your tenets are refuted.
Let me be terse:
"Our universe
   Grows daily more diluted!"

Said Hoyle, 'You quote
Lemaitre, I note,
   And Gamow. Well, forget them!
That errant gang
And their Big Bang —
   Why aid them and abet them?

'You see, my friend,
It has no end
   And there was no beginning.
As Bondi, Gold,
And I will hold
   Until our hair is thinning!'

'Not so!' cried Ryle
With rising bile
   And straining at the tether;
'Far galaxies,
Are, as one sees,
   More tightly packed together!'

'You make me boil!'
Exploded Hoyle,
   His statement rearranging.
'New matter's born
Each night and morn
   The picture is unchanging!'

'Come off it, Hoyle!'
I aim to foil
   You yet' (The fun commences)
'And in a while,'
Continued Ryle,
   'I'll bring you to your senses!'

The Lonely Life of a Double Planet

One often wonders why, despite constant 'sightings' made by cranks, no real alien spaceship from the stars has ever landed on Earth. Jerome Pearson, a senior scientist with the US Air Force, gives an astonishing but most convincing explanation:

'Where are the extraterrestrials? Why haven't they landed in their flying saucers on the White House lawn to welcome humanity to the Galactic Club?' Enrico Fermi asked this pertinent question back in 1939, long before the search for extraterrestrial intelligence (SETI) began. In the past few decades, Fermi's question has taken on a new urgency as we have built powerful radio receivers and listened to tens of thousands of ever-silent stars, and even sent our own signals to unresponding planets.

Fermi's paradox arises from a chain of sound logic. There are thousands of millions of stars in our Milky Way galaxy that are very much like the Sun, warm and long-lived. Many millions of these stars may be accompanied by planets and thousands or even millions of these planets may be suitable for life. If life arises naturally, then we might expect to find thousands of living planets and perhaps hundreds of civilizations in our Galaxy.

In the 1960s, the astronomer Frank Drake wrote an equation to estimate the number of civilizations in the Galaxy. This was based on reasonable guesses about the proportion of suitable stars and planets, and the probability that life, intelligence and civilization would develop. Any seemingly consistent set of numbers in Drake's equation predicts a multitude of life in the Galaxy. Our Galaxy is between 10 and 15 billion years old. If there are civilizations, some of them should be much older than ours. Travelling at a mere 1 per cent of the speed of light, their spaceships could have covered the entire Galaxy in a few million years. Yet the ever-expanding SETI programme has found no evidence of intelligence. There are no signs that they visited us, are on the way, or even that they are communicating with each other. Fermi's paradox has become more of a mystery than ever.

What are the possible answers? Over the past decade, scientists have considered many. Leaders of the quest for extraterrestrial intelligence, like Carl Sagan, imagine that interstellar travel is difficult or impossible, that spreading civilization between the stars is a slow and painful process, and that societies destroy themselves before they can reach us.*

Others, such as Anthony Martin and Alan Bond, see the lack of extra-terrestrials as proof that we are alone in the Galaxy, if not in the whole Universe. Many of these scientists use biology to argue that the occurrence of life by chance is so remote as to be unthinkable. Others believe that evolution is such a random process that no two intelligent species developing on different planets could have enough in common even to recognize each other, much less to communicate intelligently.

Another group denies the very existence of the Fermi paradox and insists that the Galaxy is teeming with life, just as Drake's equation predicts, and even that extraterrestrials have visited us. They have not revealed themselves because they are studying us. An interstellar civilization could know all there is to know about stars, planets and solar systems. The only unpredictable thing that could arouse their scientific curiosity would be the extraordinary variety of alien life forms, and they have therefore set us aside for observation as a 'zoo'.

Let us take a fresh look at these contradictory theories. Something is wrong either with our assumptions or with our logic. One basic assumption is that of *ordinariness*. We assume that the Earth and its life are typical, placed in a moderate location in a typical solar system around an average star. Yet the Earth is not typical at all; for several reasons it is unique in our own Solar System.

The Earth and its Moon are more like a double planet than a primary and its satellite. The Moon is far larger compared with the Earth than any other satellite of a major planet in the Solar System. Jupiter and Saturn have respectively 317 and 95 times the mass of the Earth, but their largest moons are hardly bigger than our Moon. Our large Moon has affected the Earth significantly. The ocean tides raised by the Moon had a profound effect on the evolution of crustaceans and amphibians. The emergence of tidewater zones, which alternate between flooding and drying out, perhaps even helped life to emerge on land.

A second unusual feature is the presence of Earth's atmosphere and oceans, which are unique in the Solar System. The amount of oxygen in the Earth's atmosphere is far too much for chemical equilibrium. This unstable condition is maintained by the constant action of living plants; without life, the oxygen would react with the materials on the surface and produce carbon dioxide. The excess carbon is tied up in enormous beds of limestone and rocks containing other carbonates that were laid down

*As one astronomer points out, it is hard to believe that *all* of them would have destroyed themselves before inventing space travel. See Michael Hart, 'An Explanation for the Absence of Extraterrestrials on Earth.' *Quarterly Journal of the Royal Astronomical Society,* Vol. 16, pp. 128–135, 1975.

millions of years ago. The oxygen not only supplies animals with energy so they can move and reproduce, but also creates and maintains the ozone layer that protects life from deadly ultraviolet radiation from the Sun.

The Earth also has a very powerful magnetic field. This field is much larger in proportion to its mass than that in any other planet. Its field is 100 times stronger than might be expected from statistical comparisons with the other planets. Existing theories cannot explain this, even when we take into account the heating caused by the tidal action of the Sun and the Moon.

Another unusual aspect of the Earth is its active, molten core. The energy of this core drives the tectonic plates, recycles new crust from the interior, and releases the same kind of gases from inside the Earth as those that produced the present atmosphere. This molten core is responsible for all the volcanoes and mountain ranges, for the opening and closing of ocean basins, and for the separation of the continents, which has isolated gene pools and speeded up evolution. The Indian researcher U.R. Rao has noted that a magnetic field is vital in maintaining the ozone layer to protect life from ultraviolet radiation.

All these unusual characteristics of the Earth may be due to the presence of our large Moon. According to the Australian geologist Stuart Ross Taylor, the Moon is large because it is a remnant of the small bodies that coalesced to form the planets in the cloud of debris that formed the young Earth. Normally, the larger planets grow at the expense of the smaller ones by collisions and by collecting debris from nearby orbits. The Moon may have been 'captured' by the Earth during an unusual collision in which the moon 'grazed' the Earth. Such double planets are likely to be very rare because the overwhelming majority of 'orbital encounters' result in either the complete shattering or merging of the two bodies after a collision, or a simple flyby after a miss.

For a billion years after its bizarre capture, the Moon was very close to the Earth; this proximity raised enormous tides on both bodies and heated their cores far above what would be normal. This prolonged heating continued until the action of the tides pushed the Moon to nearly its present distance, and slowed down the rotation of the Earth from about four hours to 24.

The Earth and the Moon are obviously a rare case, but how rare? It is difficult to predict the likelihood of such double planets occurring. There is admittedly some evidence in the Solar System for the existence of double planets. It consists of those miniature solar systems represented by the asteroids.

Of the thousands of known asteroids, there are as yet none confirmed to be double. Recent observations by Luke Flynn showed that none of the 17 most likely candidates, including Ceres, Pallas and Vesta, had any companions. Nor was there any evidence of dust clouds around these asteroids.

Collisions and close encounters during orbits have been common among these satellites and asteroids, as their cracked, scarred, and pitted surfaces show. But from not one of the many interactions has there formed a double satellite comparable to the Earth-Moon system. All of them produced either rings of debris or interacting bodies that keep their distance. About one tenth of the satellites have been affected; if these grazing collisions between satellites and asteroids are rare, then perhaps only one in a hundred, or one in a thousand, planets will be part of a double act.

This result alone reduces the expected number of habitable planets in the Galaxy to far below the millions foreseen by Drake. And another factor could lower the number still further. Studies of the Earth's climate show that even a small change in our average distance from the Sun could have turned the Earth into a runaway greenhouse or a cold, dry desert. The range of orbits where life is possible — the habitable zone — is small. A little closer to the Sun and we would fry; a little farther away and we would freeze. It may be that the habitable zone is far smaller than astronomers thought, or even that there is no permanently habitable zone at all.

The requirements for life can be summarized as follows: a strong magnetic field is necessary so that the planet can become 'active'; only an active planet offers protection from cosmic radiation, recycles crustal material, and produces oxygen by the release of gas from the interior. A single planet cannot develop such an internal field without being too massive or rotating too rapidly to be able to support life. Even a double planet must have developed life very quickly in order to forestall the instabilities of greenhouse effect or glaciation. All habitable planets must therefore be double, and all single planets must be uninhabitable.

This implies that there may be two types of terrestrial planets: the first type consists of double planets, with very strong magnetic fields, plate tectonics, tides, life, and stable (or moderate) climates; the second type consists of single planets, with weak magnetic fields, incomplete plate tectonics, and unstable climates that are either too hot or too cold. If this is correct, we may be alone, or one of a very few civilizations in the Milky Way.

The implications for humanity are that we may be alone — this Galaxy, at least, belongs to us. Our species may be the only opportunity for life to spread across it.

The American Indians made a very wise choice when they selected the words for treaties meant to last forever:

> *As long as the Moon shall rise,*
> *As long as the rivers shall flow,*
> *As long as the Sun shall shine,*
> *As long as the grass shall grow.*

EUREKA!

These words encapsulate the requirements for life — a double planet with a large moon, a temperate zone where water can flow and a warm and lasting Sun.*

The Cosmic Network

But suppose there are other intelligences in the universe, even if they number only a few per galaxy. Timothy Ferris proposes an ingenious way for them to communicate with one another, without having to wait thousands, or millions, of years to get a reply to a message at the limiting speed of light. And even if it turned out that we were alone in space, this system would work just as well between isolated *human* communities of the far future as it would between ourselves and aliens.

The first signal to be acquired in the future by a SETI receiver might very well have already been dispatched, not by the inhabitants of another planet, but by an intelligent computer. To see how this could be so, we need only consider the practical needs that an advanced civilization would have once it had been in the interstellar communication business for a while.

Suppose that yours is one among 101 worlds in the Milky Way galaxy that have established radio communication with one another. You now have a minimum of 100 antennae in action, each maintaining contact with a different planet thousands of light-years away. This arrangement has two drawbacks. First, it is inefficient; for the sake of economy you would prefer to use as few antennae as possible. Second, far more serious, is the Question and Answer time; if you ask a question it takes thousands of years to get an answer.

The way to alleviate both problems is to *network* the system. You install a single, automated station in space to handle *all* the radio traffic, and you link it to your planet via a single antenna system. Getting out your map of the Galaxy, you then determine strategic locations for siting other such

* In 1993, scientists at the Bureau des Longitudes in Paris discovered another reason why intelligent life might not have evolved on a moonlit Earth. Without the Moon's gravity the Earth's axial tilt, instead of varying between 22 and 25° over a period of 40,000 years, would increase and decrease by up to 50°. The result would be chaotic onsets of ice ages and epochs of drought.

86

automated stations, and you transmit a request to the worlds located at those junctures to build them. Soon — meaning in a matter of a few dozen millennia or so — everybody is sending and receiving data to and from all the other worlds through local junction terminals, which may be in their own star system or in the one next door. This way they need not employ separate antennae for each planet with which they communicate, any more than Earthlings employ a separate telephone for every person they call.

The network could have several salutary features. For one, it could be instructed to acquire signals from new worlds and bring them on-line. Indeed, its stations might broadcast signals simply to attract the attention of relatively undeveloped planets. If so, the first signal intercepted by a SETI radio telescope might come from an automated station far from any inhabited planet. To accomplish this and other tasks efficiently, the network should be able not only to repair itself but to expand as the growing body of data requires. Using self-replication, the network could send probes to strategically favourable star systems in the Galaxy, where each would build itself up into a *new* junction that would in turn hook up with the rest of the network.

Most important, the network should have a self-expanding memory that regularly updated itself at every station. It would thus alleviate the Q and A problem. What one really wants from interstellar communication is, not conversation, which takes too long, but *information*. One wants to know who else lives in the Galaxy, what they look like, how they think and what they do; and about their history and that of the species that preceded them. To make this and all other information available to everyone interested, the network should *remember* everything that it conveys.*

---

* Very different from an office computer network system on Earth, which must erase its data after it has been used through lack of storage space! But the boundless vastness of space could provide unlimited facilities for data storage. Whole asteroids, for example, could be adapted for this purpose. — A.B.

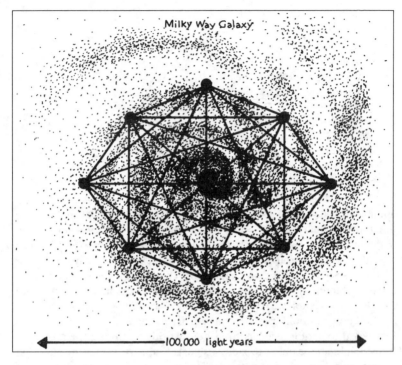

Direct radio communication between intelligent species would be relatively slow and inefficient. Here, eight inhabited worlds scattered around the galaxy are communicating directly. The average Q&A time is one hundred thousand years.

The network, then, would not only be a telephone or television system, but also a computer and a library, to which access would be as near as the nearest junction. If a species of intelligent birds on one side of the Galaxy were interested in the biology of a species of intelligent reptiles on the opposite side of the Galaxy, they would not have to send a message and wait 200,000 years for a reply. Instead the information would already be stored in the memory banks of the network itself, and the requisite Q and A time would be little longer than the light travel time to and from the nearest network junction. Nor would the information be hostage to the fate of any particular world; once submitted to the network, it would survive indefinitely.

We thus envisage an immortal system, constantly expanding and continually acquiring and storing information from all the worlds that choose to subscribe to it. In the long run, the network itself might evolve into the single most knowledgable entity in the Galaxy. It alone could survey the full sweep of Galactic history and experience the development of knowledge on a panstellar scale. Growing in sophistication and complexity with the passage of aeons, forever articulating itself among the stars, the network would come to resemble nothing so much as the central nervous system of the Milky Way.

Which, perhaps, is the ultimate purpose of intelligence, if life and intelligence may be said to have a purpose. We often find that our deepest yearnings have less to do with ourselves than with the wider scheme of things. Perhaps this is true as well of our deep but seemingly inexplicable desire to learn whether we are alone in the Galaxy. Life might be the Galaxy's way of evolving a brain.

The process could extent beyond the Galaxy, too, through contact with similar networks in other galaxies. Intergalactic Q and A times go to many millions of years — too long for mortal beings to wait, but perfectly manageable for an interstellar network. The network could afford to fashion giant antennae, use them to broadcast powerful signals to the Andromeda galaxy, even to the populous heart of the Virgo supercluster

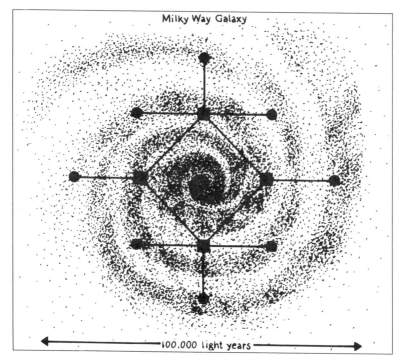

Networking of interstellar communication greatly improves the efficiency of the system. This rudimentary network, consisting of only four junctions, cuts the Q&A time (for communication with the terminals' memory banks) in half, to fifty thousand years. By adding more junction stations to the network, the Q&A time can be reduced to a century or less.

of galaxies, 60 million light-years distant, and then wait for a reply. Every world on every network would stand to benefit as galaxy after galaxy established contact, spinning electromagnetic threads across the expanding universe and exchanging the wealth of galactic libraries. The human species is only about two million years old, a time equal to that required for a message to travel one-way from the Andromeda galaxy to ours. But if information about the Andromeda galaxy and the history of its worlds were *already stored* in our Galaxy's network, we might be able to begin accessing it within a matter of decades after making contact.

All this may be a dream. Yet it points to an idealistic vision of worlds by the thousands, some in their youth and some in their graves, linked by an intergalactic mechanism devoted to pure thought. And, it suggests a cosmic role for intelligence — the combination of intelligence and technology to awaken the universe to its own life and thought and history. That would turn us all into the substance of a cosmic mind.

4
Communicating

Telegraph Hill

Before the coming of the electric telegraph, ways were invented to send complex messages accurately at great distances at comparatively high speeds. Anthony Michaelis explains:

We would probably still live in caves had not people learned to communicate and pass on to their children the knowledge they had gained in their short and dangerous lives. With writing, it became possible to communicate over a distance, both in space and time, and we would know but little of our early ancestors, had they not left their inscriptions in stone, clay and metal, wood, paper and silk.

For millennia the speed of communications remained that of the swiftest runner or the fastest horse, perhaps a distance of 10 miles an hour. (Greek and Roman signal fires and African drums were of course somewhat faster.) Only with the coming of the telescope was there any hope of swift and accurate communication over long distances.

It was the 17th century English physicist Robert Hooke who first gave an outline of visual telegraphy in 1684. But his system was never tried out. Over a century later, a brilliant French engineer Claude Chappe again took up the challenge. He produced a practical system which could send messages all over France. When it was finally superseded in 1852 by the electric telegraph, France was covered by a network of 556 sempahore stations stretching over a total distance of 3,000 miles.

During the Revolution, France had a desperate need for swift and reliable communications. The country was surrounded by the allied forces of Britain, Holland, Prussia, Austria and Spain. Marseilles and Lyons were in revolt, and the British Fleet held Toulon. The only French advantage was the lack of co-operation between the allied forces, caused by their bad communications.

Chappe and his brothers in 1790 set out to devise a communication system that would allow the central government to receive and transmit in the shortest possible time. He carried out his experiments during the next two years, and twice his apparatus at the Etoile in Paris was destroyed by the furious mob who thought he was communicating with the imprisoned King, Louis XVI. His efforts were recognized in 1793, when he was appointed *Ingénieur-Telegraphiste* and ordered to establish a line of telegraph stations between Paris and Lille, a distance of 140 miles.

His stations were simply towers, old towers or towers specially built for this purpose. On each of their roofs was a vertical wooden extension to which was fixed a movable horizontal beam, which could be swung into various angles by means of ropes. At the end of each beam were two moveable vertical arms. Many coded configurations could thus be achieved, and these could be read through a telescope from the next tower.* The first message to reach the government by these means was that their forces had retaken Le Quesnoy. A fortnight later, another message was joyfully received in Paris, reporting the recapture of Condé.

No wonder then that the telegraph was extended throughout France. Paris to Strasbourg with 50 stations was the next line, and others soon followed. But because each station had to be within sight of the next, the cost of administration and wages were huge. Only when the telegraph was linked with a lottery did they drop. Chappe killed himself in 1805, when the anxiety became too great for him.

The reports of his telegraph reached England in 1794, and stimulated Lord George Murray to propose a similar system to the British Admiralty. On *his* towers, Murray used a large wooden board. Each board had six large circular holes which could be closed by wooden shutters. A chain of these stations, 15 in all, was erected between London and Deal at a cost of nearly £4,000; others followed to Portsmouth, Yarmouth and Plymouth. The line to Portsmouth was not finally closed down until 1847, and some of the hills on which the towers were built are still today known as 'Telegraph Hill'.

The visual telegraph was the fastest means of communication at the time. To quote from a contemporary: 'A single signal has been transmitted to Plymouth and back to London in three minutes, which by the telegraph route is at least 500 miles. The progress was at the rate of 170 miles in a minute, or three miles per second, or three seconds at each station — a rapidity truly wonderful!' But the system had one serious disadvantage. Night and bad weather brought it to a halt. 'The station on Putney Heath communicating with Chelsea,' said this report, 'is generally rendered useless during easterly winds by the smoke of London which fills the valley of the Thames.'

The history of the electric telegraph, which superseded visual signalling, began on February 17, 1753, when a letter signed by a certain C.M. was

---

* This system was used by the hero of Dumas's novel *The Count of Monte Cristo* to send false financial messages to his enemies.

published in the *Scots Magazine*. His identity has never been established. He proposed that 'a set of wires equal in number to the letters of the alphabet be extended horizontally between two given places, parallel to one another and each about an inch apart.' They would signal a particular letter through an electrostatic machine.

It had been known since very early times that electrostatic forces would attract small pieces of paper, and by the middle of the 18th century simple frictional machines could produce electrostatic energy. C.M. proposed to use the electricity from such a machine, channel it through one of his wires, and let it attract on the receiving side the corresponding piece of paper with its letter of the alphabet.

But 1753 was hardly a suitable date for electrical telegraphy. Static electricity was then more often used to entertain the 'philosophical' friends of the owner of a frictional machine. It was common among them to transmit an electric shock through a circle of twenty or thirty people, each holding hands, so that all would experience the shock simulta-neously.*

However fast the speed of transmission of electricity, it was a long time before it was taken seriously.

As early as 1787, a Spaniard named Betancourt carried out experiments with static electricity to send messages between Madrid and Aranjuez, a distance of 25 miles. Another experimenter was Sir Francis Ronalds, an

* One such pointless experiment was performed on a grand scale by the Abbé Nollet (1700–70.) He passed a shock round a circle of Carthusian monks nearly a mile in circumference, whom he had linked together by strands of iron wire.

English merchant, who began to experiment in static electricity in 1816. To demonstrate the speed of electric transmission, Sir Francis erected in his London garden two large wooden frames with 26 parallel wires, to control each letter of the alphabet, suspended between them. An electric signal sent along a given wire indicated which letter was being transmitted. (Another 26 wires went the other way, so that the operator at the other end could send a reply.) It was a highly ingenious system and it worked.

It certainly deserved serious official consideration, but when Sir Francis submitted it to the Admiralty, he was informed that 'telegraphs of any kind are wholly unnecessary.' No doubt Lord George Murray had better friends at the Admiralty than he did.

This brief history of the precursors of the electric telegraph demonstrates a basic truth about technology: an inventor needs both to develop and exploit his schemes before he can expect either acknowledgment or reward. And if they are too far ahead of the technical capabilities of his age, then all he can hope for is an honourable mention in the history books of his science. Only very rarely can one person's unsuccessful ideas be directly traced down to the pioneer who brought them to fruition.

The Conquest of Solitude

Telephones used to be called 'Bell telephones'. This was not because they ring, as some people imagine, but in honour of their inventor, Alexander Graham Bell.

Bell, a native of Edinburgh and a naturalized American, was in love with a deaf girl and wanted to invent machines that would help the deaf. As professor of vocal physiology at Boston University, he became engrossed in the idea of reproducing sounds mechanically. It seemed to him that if sound waves could be turned into electric current, then the current could be converted back into sound at the other end of a circuit.

He invented the telephone quite by accident. One day in 1876 he spilled battery acid on his trousers while experimenting with an instrument designed to carry sound. He called for his assistant Thomas Watson without realizing that Watson was in another part of the building. He described in his laboratory notes what then happened:

I had shouted into the mouthpiece the following sentence: 'Mr. Watson, come here. I want to see you.' To my delight he came, and declared that he had heard and understood what I had said. I asked him to repeat my words. He did so exactly. We then changed places. I listened at the receiver while Watson read a few passages from a book into the mouthpiece. It was certainly the case that articulate sounds proceeded from the receiver. The effect was loud, but indistinct and muffled.

I could not make out the sense, but an occasional word here and there was quite distinct. I made out 'to' and 'out' and 'further'; and finally the sentence: 'Mr. Bell, do you understand what I say?' came out quite clearly and intelligibly.

Bell patented the telephone that same year. A few months later, it was the great hit of the centenary celebrations of the Declaration of Independence. The visiting Brazilian emperor Pedro II made world headlines by exclaiming, 'It talks!'

Yet people did not immediately foresee today's vast proliferation of telephones. An American mayor was thought most bold when he predicted, 'One day there'll be one in every city.' And in Britain the chief engineer of the Post Office, Sir William Preece, told a Commons committee, 'The Americans have need of the telephone — but we do not. We have plenty of messenger boys.' Arthur C. Clarke has since forecast that before the end of the twentieth century there will be a telephone, not just in every village on Earth, but in every home.

Hearing Voices

Thomas Edison improved the telephone and invented the phonograph, the precursor to the gramophone. It did not get easy acceptance, writes Joe Nickell:

In 1878, members of the French Academy of Sciences had gathered to witness the demonstration, by the physicist Du Moncel, of Edison's recent invention. The meeting was attended by Jean Bouilland, a celebrated physician. As the small, primitive phonograph began to speak — faithfully reproducing the words Du Moncel had spoken just moments before — suddenly the 82 year-old Bouilland leaped at the physicist, grabbing him by the throat.

'You wretch!' he shouted. 'How dare you try to deceive us with the ridiculous tricks of a ventriloquist!' Bouilland 'knew' that only people — and not machines — could speak.

Stop Screwing and Answer That Phone!

The telephone [Dennis Flanagan relates] is now so embedded in human behaviour that it is hard to imagine what life was like without it. It has shaped human behaviour. Many Europeans have observed that Americans seem to have an overpowering urge to answer the telephone, even in the midst of more important business. This behaviour may have a distinct historical origin.

The cost of a telephone call — not the charge to the customer, but the cost to the telephone company — depends partly on the length of the call. In order to handle x calls with an average length of y the company needs z facilities. Therefore if y is longer, the company needs more z, and if y is shorter, the company needs less.

In the First World War, analysts at the Bell system realized that if people would learn to answer the telephone a little faster, the average length of the calls would be shorter and less facility would be needed. These savings, multiplied by the company's vast number of facilities, would add up to a substantial sum.

The company launched a vigorous advertising campaign urging people to answer the telephone faster. The appeal was put on the basis that tying up telephone communications hindered the war effort. It worked like a charm. The man in charge of it later remarked: 'I have made the Americans the only people in the world who will interrupt sex to answer the telephone.'

Glossary of Incompetence

Authoritative-sounding statements in scientific journals should not always be taken literally. I.J. Good has made a collection of them:

'*It has long been known that . . .*'
I haven't bothered to look up the original reference.

'While it has not been possible to provide definite answers to these questions . . .'
The experiment didn't work out, but I figured I could at least get publication out of it.

'High purity . . .' 'Very high purity . . .' 'Extremely high purity . . .' 'Super purity . . .'
Composition unknown except for the exaggerated claim of the suppliers.

'. . .accidentally strained during mounting,'
. . . dropped on the floor.

'It is clear that much additional work will be required before a complete understanding . . .'
I don't understand it.

'Unfortunately, a quantitative theory to account for these effects has not been formulated . . .'
Neither does anybody else.

'It is hoped that this work will stimulate further work in the field.'
This paper isn't very good, but neither is any of the others on this miserable subject.

'The agreement with the predicted curve is excellent.' '. . . good.' '. . .satisfactory.' '. . .' fair.'
Fair. Poor. Doubtful. Imaginary.

'As good as could be expected considering the approximations made in the analysis.'
Non-existent.

'Of great theoretical and practical importance.'
Interesting to me.

Three of the samples were chosen for detailed study.'
The results on the others didn't make sense and were ignored.

'These results will be reported at a later date.'
I might possibly get round to this some time.

'Typical results are shown.'
The best results are shown.

'Although some detail has been lost in reproduction, it is clear from the original micrograph that . . .'
It is impossible to tell from the micrograph.

'It is suggested . . .' 'It may be believed . . .' 'It may be that . . .'
I think.

'*The most reliable values are those of Jones.*'
He was a student of mine.

'*It is generally believed that . . .*'
A couple of other guys think so too.

'*It might be argued that . . .*'
I have such a good answer to this objection that I shall not raise it.

'*Correct within an order of magnitude.*'
Wrong.

'*Well known.*'
(i) I happen to know it; (ii) well known to some of us.

'*Obvious,*' '*of course.*'
(i) I was not the first to think of it; (ii) I also thought of it independently, I think.

Death by Secret Writing

Mary, Queen of Scots, executed at Fotheringay Castle, Northamptonshire, on 8 February 1587, died from scientific incompetence. She used ciphers without understanding the first rule of secret writing: that a cipher must be sufficiently secure, so that enemies cannot crack it.

Mary used a substitution cipher, in which each letter of the plaintext message is replaced by a symbol previously agreed with one's correspondent. It was essentially the same system that Julius Caesar had used, advancing each letter one place in the alphabet, so that the words DEAR CICERO would be enciphered as EFBS DJDFSP.*

But the art of code-breaking had progressed since messengers bore Caesar's ciphertexts through lands peopled by illiterate barbarians. Substitution ciphers could now be cracked. It proved to be a simple matter of knowing the frequency in which letters tend to occur in a given language. In English, for example, as every good Scrabble player knows, the commonest letter is E, followed by T, A,

---

* On reaching the end of the alphabet, one would simply start again, so that Z would be enciphered as A.

O, I, N, S and R. To crack the ciphertext — provided it is sufficiently long — the code-breaker finds the most frequently occurring symbol, which must be E. The next most frequent will surely be T, and so on.*

The imprisoned Queen of Scots was a passionate user of ciphers. She smuggled out hundreds of cryptograms to Queen Elizabeth's enemies in The Hague and Madrid. However, the danger of frequently recurring symbols being noticed by an enemy never occurred to her, and she died, not so much by the headsman's axe, as from the disease of bad secret writing.

Her death was science's great lesson to diplomats. Never again did governments use simple substitution ciphers.** In fact, only a few months before her execution, a French nobleman, Blaise de Vigenère, invented a cipher that was far more difficult to crack. It used 26 different alphabets simultaneously, so that the encipherment of plaintext letters was usually different each time. In one place E might be translated into X, in another to B. Vigenère ciphers were used in different variations right up to the Second World War, when the last one ever put to serious use, the highest-security German Enigma code, was cracked by British code-breakers at Bletchley Park with the use of a primitive computer.

The story of Mary's avoidable death is told by David Kahn in his book *The Codebreakers*:

King Phillip II of Spain ruled the Netherlands through its governor, his half-brother Don John of Austria. Don John dreamed of crossing the Channel into England with a body of troops and dethroning Elizabeth.*** He would then marry the Queen of Scots and share with her the Catholic crown of England. Philip secretly consented to the plan.

But England did not sleep. Sir Francis Walsingham, Elizabeth's satanic-looking minister, had built up an efficient organization for secret intelligence. He had a bright young man who could devour enciphered messages. This was Thomas Phelippes, England's first great cryptanalyst.

---

* The remaining letters in order of frequency in English are H, L, D, C, U, M, F, P, G, W, Y, B, V, K, X, J, Q and Z. For frequency order of *words*, see the Appendices to Caxton Foster's fascinating book *Cryptanalysis for Microcomputers*.

** They were still being used *in fiction* right up to the twentieth century. Edgar Allan Poe's short story *The Gold Bug* and the Sherlock Holmes story *The Dancing Men* both involve substitution ciphers. In the latter the exposed villain exclaims to Holmes: 'There was no one on earth outside the Joint who knew the secret of the dancing men. How came you to write in it?' He ought to have read about the Queen of Scots.

*** The romantic Don John was considered the foremost warrior in Europe on account of his defeat of a Turkish fleet at the battle of Lepanto in 1571. G.K. Chesterton made this victory the subject of a heroic poem.

Phelippes was one of the minister's most confidential assistants. He was an indefatigable worker. He appears to have been able to solve ciphers in Latin, French and Italian. The only known physical description of him comes from the pen of Mary herself, who describes him as 'of low stature, and eated in the face with smallpox.'

Mary's unflattering comments betrayed her suspicions about Phelippes — suspicions that were well founded. For Phelippes and Walsingham were casting a jaundiced eye on Mary for reasons that were equally well founded. Catholic factions had schemed more than once to seat her on the throne of England and so restore the realm to their Church. The chief result of these intrigues was to alert Walsingham to seek the chance to extirpate once and for all this cancer that threatened his own queen, Elizabeth.

The opportunity arose in 1586. A former page of Mary's, Anthony Babington, began organizing a plot to have Elizabeth assassinated, incite a Catholic uprising, and crown Mary queen.* Babington gained the support of Philip II, who promised to send an expedition. But the plan depended on Mary's acquiescence, and to obtain this Babington had to communicate with her.

This seemed no easy task. Mary was then being held incommunicado at Chartley Castle in Staffordshire. But a former seminarian named Gilbert Gifford, recruited by Babington as a messenger, found a way to smuggle letters to Mary into the castle in a beer keg.

Many of these letters were enciphered. But that was only part of the care that Mary took to ensure the security of her communications. She insisted that important letters be written in her suite and read to her before they were enciphered. She not infrequently ordered changes in her substitutions.

What neither she nor Babington knew was that their correspondence was being delivered to Phelippes and Walsingham as quickly as they wrote it. For Gifford was a double agent. Walsingham had seen an unparallelled chance to spy on Mary. He employed Gifford to give him all Mary's letters, which he copied before passing them on. They included the rapidly growing traffic generated by Babington's plot. Phelippes solved these enciphered missives almost as soon as he got his hands on them.

Walsingham wisely made no arrests, but let the plot develop and the correspondence accumulate in the hope that Mary would incriminate herself. His expectations were soon fulfilled. In July, 1586, Babington specified the details of the plan in a letter to Mary, referring to Don John's intended invasion, her own deliverance, and 'the dispatch of the usurping competitor' (Elizabeth).

___

* The Babington plot should in reality be called Walsingham's plot. Many of Babington's fellow-conspirators were government spies. Walsingham secretly encouraged the plot, wanting a legal pretext for Mary's execution.

Mary considered her reply for a week, and after composing it carefully, had it enciphered and sent to Babington. It was to prove fatal. For in it, Mary acknowledged 'this enterprise' and advised Babington of ways 'to bring it to good success.' Phelippes, on solving it, triumphantly drew a gallows on his copy of the letter.

But Walsingham still lacked the names of the six young courtiers who were going to commit the actual assassination. So when the letter reached Babington, it bore a postcript *that was not on it when it left Mary's hands.* In it, Babington was asked for 'the names and qualities of the six gentlemen which are to accomplish the designment.' The added forgery, in Mary's cipher, was the work of Phelippes.

Her letters served as thoroughly incriminating evidence in the Star Chamber trial that convicted her of high treason. She received the announcement that Elizabeth had signed her death warrant with majestic tranquillity. At eight in the morning of February 8, 1587, she mounted the scaffold and received the executioner's three strokes. Thus died Mary Queen of Scots. There is little doubt that bad secret writing hastened her end.

The Unbreakable Cipher

He spent part of the night working carefully with his one-time pads, preparing a coded message that no computer could break.

Frederick Forsyth, *The Fourth Protocol*

It used to be believed that, no matter how cunningly a system of secret writing was designed, there would always be someone still more cunning who could unravel it.

This idea was demolished for ever in 1917. Gilbert Vernam, an employee of American Telephone and Telegraph, was instructed to produce a method of secure communication which the Germans could not crack. He invented what is now called the One-Time Pad.

His idea was astonishingly simple, and it seems incredible in retrospect that so many generations of diplomats and spies did not themselves discover it.

It may have been because they lacked a simple and widely known code like Morse, or ASCII, upon which to attach a second, and impenetrable encipherment.

I must explain this briefly. The ASCII code (the initials are short for 'American Standard Code for Information Interchange') is a device invented by programmers to identify each key on a standard computer keyboard with an individual number. It is quite arbitrary and illogical. The capital A, for example, has the number 65, and B is 66. The capital letters A to Z thus go from 65 to 90. All the punctuation signs have their own numbers. Even the spacebar — creating the spaces betwen words — has its own number, 32. The lower-case letters have different numbers, as do the numerals. For example, *d* is 100, the numeral 3 is 51 and the dollar sign is 36. (The full ASCII code, which is the same for all computers, whether mainframe, mini or micro, can be found in any popular book about computers.)

Anyway, Vernam's unbreakable cipher goes like this: Plaintext, the message or document to be encrypted, is translated into its ASCII equivalent. Each number in the ASCII version is then *added* to the numbers in a secret key, which each correspondent uses once and only once.

A well-equipped spy — or an embassy — will use a 'book of pads', a book of many pages consisting of columns of random numbers. His correspondent will possess the only other identical copy of this book. The agent uses each page of the book to encrypt one message, and then destroys that page.

Here is an example of how it might work:

In Encryption:

PLAINTEXT	T	H	I	S		I	S		S	E	C	R	E	T
ASCII CODE	84	72	73	83	32	73	83	32	83	69	67	82	69	84
SECRET KEY	31	17	9	81	14	2	56	8	17	23	66	14	7	44
ADDITION	51	89	82	100	46	75	75	40	36	92	69	32	76	64
CIPHERTEXT	3	Y	R	d	.	K	K	(	$	\	E		L	@

(To decrypt, one merely performs the same operation in reverse, turning the ciphertext into ASCII, and then *subtracting* the secret key from the answer to that sum.) 3YrD.KK($\E L@ is thus turned back to THIS IS SECRET. Part of the encryption process may seem somewhat mysterious. Some of the additions seem at first sight to be wrong. In the case of the first S, why does 83 and 81 add up to only 100? The answer is, that in this case, for simplicity, we are using a restricted range of ASCII numbers. The rule here is that whenever an addition of the ASCII number and secret key is greater than 95 we automatically subtract 64. Thus in the case of that first S, $83+81-64=100$. It does not *have* to be done like this, but it makes presentation simpler. (It also has the minor disadvantage of restricting plaintexts to capital letters.)

So why is the one-time cipher unbreakable? David Kahn explains:

The one-time cipher consists of a random key used once, and only once. It provides a new and unpredictable key character for each plaintext character in the whole ensemble of messages ever to be sent by a group of correspondents.

It is an unbreakable system. Some systems are unbreakable in practice only, because the cryptanalyst can conceive of ways of solving them if he had enough text and enough time. The one-time cipher is unbreakable both in theory and in practice. No matter how much text a cryptanalyst had available in it, or how much time he had to work on it, he could never solve it. This is why:

To solve the cipher, one must gather all the enciphered letters into a homogeneous group that may be studied for its linguistic traits. But the cryptanalyst has no way of sorting them out because the key in a one-time system neither repeats, nor recurs, nor makes sense. Hence his methods all fail. The perfect randomness of the system and its one-time nature nullifies any cohesion, as in keys that are repeated in a single message or in several messages. The cryptanalyst is blocked.

How about trial and error? It might seem that brute testing of all keys, one after another, would eventually yield the plaintext. But success this way is an illusion. For while exhaustive trials would indeed bring out the true plaintext, they would also produce every other possible text of the same length, and there would be no way to tell which was the right one.

Suppose that the cryptanalyst deciphers a four-letter military message with every key he can think of, beginning with AAAA. He finds a plaintext that means something with AABI: *kiss*. Unlikely in this context. He presses on. The key AAEL yields plaintext *kill*. This is better — but he wants to make sure. He continues through AAEM, giving *kilt*, which might refer to some Scottish manoeuvre, and AAER, *kiln*. Further down the list he reaches *fast* at GZBM and *slow* at KHIA, *stop* at HRIW and *gogo* at XSTT, *hard* at PZVQ and *easy* at RZBU. He finds when he ends at ZZZZ that he has merely compiled a list of *every possible four-letter word* — the hard way. He can no more pick the right solution from this list than he can from a dictionary of military terms.

The key does not help in limiting the selection because, since it is random, any group of four letter words is as acceptable as a keytext as any other. The worst of it is that the possible solutions increase as the message lengthens. There are only three possible solutions for a one-letter cryptogram, but dozens for those of two letters, and zillions for those of 100.

A final hope flickers. Suppose that the cryptanalyst obtains the key to a given cryptogram, perhaps through theft or the error of a radio operator. Can he use that key to determine the underlying system on which it was built, and so predict future keys? No, because a random key has no underlying system — if it did, it would not be random.

These are empirical proofs. It is possible, however, to demonstrate a proof that the one-time system is theoretically unbreakable.

The Vernam encipherment constitutes an addition — an addition based on the Morse, or ASCII alphabet, but an addition nonetheless. Suppose than that the plaintext is 4 and the key is 5. The ciphertext will be 9. Now, given only this, the cryptanalyst has no way of knowing whether it results from the addition of $7+2$, or $6+3$, or $-2+11$, or $4+5$, or any of the other 32 possible combinations. Generalized, the situation is $x+y=9$. Mathematicians call this an equation with two unknowns, and a single such equation has no unique solution. *Two* equations with the same two unknowns are required.*

The one-time system prevents the cryptanalyst from ever bringing two or more such equations together. The utter absence of any pattern within its key precludes him from finding two occurrences of a given key character by reconstructing a pattern. And its exhaustless novelty makes it impossible for him to locate these occurrences in any key repetition.

He is thus denied any additional information to delimit one of the unknowns; he is left with all 32 possibilities for each key character (all the capital letters of the alphabet plus the punctuation signs), and consequently all 32 of the plaintext. It is true that some solutions are more probable than others. Thus, there is a 12 per cent chance that the unknown character of ciphertext is E, an 8 per cent chance that it is T, and so on down the frequency table. But this does not solve the problem, for it does not specify which of these probabilities is actually present in the individual case before him.

So the answers again evade him. Formless, endless, the random one-time cipher vanquishes him by dissolving into chaos on the one hand and infinity on the other. The cryptanalyst gropes through caverns measureless

---

* These, we were taught at school, are called 'simultaneous equations'. — A.B.

to man. His quest is Faustian; who would dare it would know more than can be known.

Now, with microcomputers, anyone can create and use their own one-time cipher. Here is a simple program in BASIC, for an IBM personal computer, written by Thomas A. Dwyer and Margot Critchfield. The program is particularly ingenious, since when the user enters his secret key it generates its own numerical key string. The program does not remove the need for each correspondent to possess his own book of secret keys; it simply makes encryption and decryption far easier than performing the task by hand:

```
10 REM ********************************************
20 REM DATA ENCRYPTION USING A ONE-TIME PAD,
30 REM BY THOMAS A. DWYER AND MARGOT CRITCHFIELD
(C) 1984
40 REM ********************************************
50 CLS: CLEAR: KEY OFF
60 DEFINT A-Z: ON ERROR GOTO 1080
70 KY$="": TX$="": REM STRING VARIABLES TO HOLD
KEY AND TEXT
80 DIM KC (255), TC (255): REM ARRAYS FOR ASCII, KEY
AND TEXT
90 INPUT "WHAT IS THE NAME OF YOUR PLAINTEXT FILE";
PT$
100 INPUT "WHAT IS THE NAME OF YOUR ENCRYPTED
FILE"; EN$
110 REM ********************************************
120 REM MENU SELECTION
130 REM ********************************************
140 PRINT: PRINT"***** MAIN MENU *****": PRINT
150 PRINT"1. TO ENCRYPT A PLAINTEXT FILE
160 PRINT"2. TO DECIPHER AN ECRYPTED FILE
170 PRINT"3. TO QUIT THE PROGRAM
180 PRINT: PRINT "ENTER YOUR CHOICE";: INPUT C
190 PRINT "PLEASE TYPE 1,2 OR 3"
200 ON C GOTO 230, 480, 690
210 GOTO 180
220 REM ********************************************
230 PRINT: PRINT" ENCRYPTING MODE
240 PRINT" ENTER YOUR SECRET KEY (UP TO 255 CHAR-
ACTERS)
250 LINE INPUT KY$:LK=LEN(KY$)
```

105

```
260 GOSUB 1010
270 REM *************************************************
280 OPEN "I",1,PT$
290 OPEN "0",2,EN$
300 IF EOF(1) THEN 420: REM LOOP THROUGH PLAINTEXT
FILE
310 LINE INPUT #1,TX$
320 LT=LEN (TX$)
330 PRINT TX$
340 IF TX$=" " THEN 390
350 GOSUB 970: REM GENERATE PAD CODES
360 GOSUB 730: REM CHANGE TEXT TO ASCII AND STORE
IT
370 GOSUB 830: REM ADD KEY CODES TO TEXT CODES
380 GOSUB 780: REM CONVERT ASCII TO CIPHERTEXT
390 PRINT #2,TX$
400 PRINT TX$: PRINT
410 GOTO 300
420 CLOSE:PRINT:GOTO 140
430 PRINT
440 GOTO 140
450 REM *************************************************
460 REM DECRYPTING MODE
470 REM *************************************************
480 PRINT:PRINT"DECRYPTING MODE
490 OPEN "I",1,EN$
500 OPEN "0",2,PT$
510 PRINT"ENTER YOUR SECRET KEY (UP TO 255 CHAR-
ACTERS)
520 LINE INPUT KY$
530 LK=LEN (KY$)
540 GOSUB 1010
550 IF EOF(1) THEN 660:REM LOOP THROUGH ENCRYPTED
FILE
560 LINE INPUT #1,TX$
570 LT=LEN(TX$)
580 PRINT TX$: IF TX$=" " THEN 630
590 GOSUB 960: REM GENERATE PAD CODES
600 GOSUB 730: REM CHANGE CIPHERTEXT TO ASCII AND
STORE IT
610 GOSUB 880: REM SUBTRACT KEY CODES FROM
CIPHERTEXT
620 GOSUB 780: REM CONVERT ASCII TO PLAINTEXT
630 PRINT #2,TX$
```

```
640 PRINT TX$:PRINT
650 GOTO 550
660 CLOSE
670 PRINT
680 GOTO 140
690 CLS: SYSTEM: REM QUIT PROGRAM
700 REM ******************************************
710 REM SUBROUTINES
720 REM ******************************************
730 FOR K=1 TO LEN (TX$)
740 TC (K) = ASC (MID$ (TX$,K,1))
750 IF TC (K) >95 THEN TC(K)=TC(K)-32
760 NEXT K
770 RETURN
780 TX$=" "
790 FOR K=1 TO LT
800 TX$=TX$+CHR$ (TC(K))
810 NEXT K
820 RETURN
830 FOR M=1 TO LT
840 TC(M)=TC(M)+KC(M)
850 IF TC(M)>95 THEN TC(M)=TC(M)-64
860 NEXT M
870 RETURN
880 FOR M=1 TO LT
890 TC(M)=TC(M)-KC(M)
900 IF TC(M)<32 THEN TC(M)=TC(M)+64
910 NEXT M
920 RETURN
930 REM ******************************************
940 REM SUBROUTINES TO GENERATE PAD KEYS
950 REM ******************************************
960 REM
970 FOR J=1 TO LT
980 KC(J)=64*RND
990 NEXT J
1000 RETURN
1010 S=0
1020 XX=RND(-1)
1030 FOR J=1 TO LK
1040 S=S+J*ASC(MID$(KY$,J,1))
1050 NEXT J
1060 RANDOMIZE S
1070 RETURN
```

```
1080 REM ********************************************
1090 REM ERROR CHECKING
1100 REM ********************************************
1110 IF ERR<>53 THEN PRINT "PROGRAM ERROR": CLO-
SE:STOP
1120 PRINT "INPUT FILE DOES NOT EXIST."
1130 CLOSE
1140 RESUME 90
```

And here is the program at work, encrypting a short passage from *Alice in Wonderland*:

> There was a table set out under a tree in front of the house, and the March Hare and the Hatter were having tea at it: a Dormouse was sitting between them, fast asleep, and the other two were using it as a cushion resting their elbows on it, and talking over its head.
> "Very uncomfortable for the Dormouse," thought Alice; "only as it's asleep, I suppose it doesn't mind."

The key is: VERNAM INVENTED THIS CIPHER

```
 ,<$U=5<A2^–F('<UL&,TE)]"*$4P^ (IL)1$VP.$=D.!
5AXD(U/H5/N—D3WC31,—ZAWF-LHJ5&%GT \&EG+
MEDPG<\&.SH[N—ZRFS%QLP=>T:G/MIIV1RHC"
(&2R.P*]5=QF*PFULQZDH9 (P' <J;PB37[OR<GOGBE <) ')
VN,#G ,HXM@&%BG ôE[GR1#[5*;(F5=A.!I/?<2GR5]
USI]71*,*!—6N@Z4 8GPP87/( R]&R)#IG[N9 'ôN@
FFBTA1W=BP++UB=N G5I1W#,@/ZT 9@S(J>WH1—
OU4/SCQ30:8T[0%80/+L#F@JIV/Y<*P2D1HF[9GAYO
E-43W*>@V4CH:#J- DZO#<PF4)X:ô2 —;FR%&RZ;
```

The Badge of Cain

The discoverer of the uniqueness of human fingerprints was a British official in India, Sir William Herschel (no relation to the great astronomer), who wrote this letter to the Inspector General of Prisons of Bengal.*

---

* A British doctor working in a Tokyo hospital, Henry Faulds, claimed to have discovered fingerprints at about the same time. But the letter in which he made this claim, published in *Nature* on 28 October, 1880, showed that, unlike Herschel, he had failed to grasp the idea of the *uniqueness* of fingerprints. — A.B.

Hooghly, August 15, 1877

My dear Bourdillon,
I enclose a paper which looks unusual, but which I hope has some value. It exhibits a method of identification of persons, which, with ordinary care in execution, and with judicial care in the scrutiny, is, I can now say, for all practical purposes, far more infallible than photography. It consists of taking a seal-like impression, in common seal ink, of the markings of the skin of the two forefingers of the right hand, these two being taken for convenience only.

I am able to say that these marks do not (bar accidents) change in the course of ten or fifteen years so much as to affect the utility of the test.

The process of taking the impression is hardly more difficult than that of making a stamp of an office seal. I have been trying it in the Jail and in the Registering Office and among pensioners here for some months past. I have purposely taken no particular pains in explaining the process, beyond once showing how it is done, and once or twice visiting the office, inspecting the 'finger signatures,' and asking the clerks to be a little more careful. The articles necessary are such that the man in charge of stationery can prepare on a mere verbal explanation.

Every person who now registers a document at Hooghly has to sign his 'sign-manual.' None has offered the smallest objection, and I believe that the practice, if generally adopted, will put an end to all attempts at impersonation.

The cogency of the evidence is admitted by everyone who takes the trouble to compare a few 'finger signatures' together, and to try making a few himself. I have taken thousands now in the course of the last twenty years, and (bar smudges and accidents which are rarely bad enough to be fatal) I am prepared to answer for the identity of every person whose 'finger signature' I can now produce if confronted with him.

As an instance of the value of the thing, I might suggest that if Roger Tichborne had given his 'finger signature' on entering the Army, the whole Orton case would have been knocked on the head in ten minutes by requiring Orton to make his 'finger signature' alongside it for comparison.*

I send this specimen to you because I believe that identification is by no means the unnecessary thing in jails which one might presume it should be. I don't think I need dilate on that point. Here is the means of

---

* Between 1866 to 1874, all London had discussed the notorious Tichborne case. Roger Tichborne, heir to a huge fortune, had been lost at sea in 1854. In 1866 a man from Australia succeeded in convincing Roger's half-blind mother that he was her long-lost son. He likewise persuaded other members of the Tichborne family and several prominent lawyers. The case dragged on for years and cost huge sums in legal fees. At length the claimant, a swindler named Orton, was convicted and sentenced to 14 years penal servitude.

identifying the identity of every man in jail with the man sentenced by the court, at any moment, day or night. Call the number up and make him stop. If it is he, it is he; if not, he is exposed on the spot. Is No. 1302 really dead, and is that his corpse or a sham one? The corpse has two fingers that will answer the question at once. Is this man brought into jail the same man that was sentenced by the magistrate? The 'finger signature' on the back of the magistrate's warrant is there to testify.

For uses in other departments and transactions, especially among illiterate people, it is available with such ease that I quite think its general use would be a substantial contribution towards public morality. Now that it is pretty well known here, I do not believe the man lives who would dare to attempt impersonation before the Registrar here. The solicitors all know the potency of the evidence too well.

Will you kindly give the matter a little patient attention, and then let me ask whether you would let me try it in other jails?

The impressions will, I doubt not, explain themselves to you without more words. I will say that perhaps in a small proportion of the cases that might come into question the study of the seals by an expert might be advisable, but that in most cases any man of judgment giving his attention to it cannot fail to pronounce right. I have never seen any two signatures about which I remained in doubt after sufficient care.

Kindly keep the specimens carefully.

Yours sincerely,

W. HERSCHEL.

Postcript:

No immediate action was taken. The recipient of this letter, Sir James Bourdillon, told Herschel years later that he had intended to do something about the matter, but he had let it 'slip through his fingers.'

Get Rid of those Knobs and Levers!

The first working computer was fundamentally a different kind of machine from any previously built. Ed Regis explains why.

In 1944, John von Neumann was planning to build his own computer at the Institute for Advanced Study at Princeton. First on the agenda, though, was to get a clear picture of the drawbacks of the ENIAC, the Electronic Numerical Integrator and Computer. Of these there was no shortage. For one thing, it was too big. In fact it was worse than big, it was colossal, a veritable dinosaur of tubes and wiring. It was 100 feet long, 10 feet high, and 3 feet deep. It had over 100,000 parts, including 18,000 vacuum tubes, 1,500 relays, 70,000 resistors, 10,000 capacitors, and 6,000 toggle switches. There seemed to be no end to the thing, and von Neumann used to joke that just keeping it going was 'like fighting the Battle of the Bulge every day.' When the machine once ran for five days without a single tube failing, the inventors were in hog heaven.

The ENIAC consumed so much power that, according to legend, every time it was turned on, the lights dimmed all over the town. From a functional standpoint, though, the machine's size, failure rate and power requirements were as nothing compared to the demands imposed by its relatively hard-wired programming. Unlike modern general-purpose computers which can switch from word processing, to graphics, to game-playing at the flick of a floppy disk, the ENIAC had been designed primarily to do one thing, and that was to compute firing and bombing tables for the Army. Getting it to do anything else was a major production. Every time you wanted to give the machine a new type of problem, you had to go

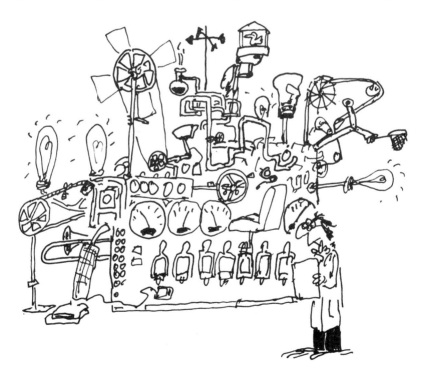

round resetting switches and replugging cables one at a time, all by hand. Since the machines had thousands of individual switches and hundreds of external cables and plugs, it could take up to two or three days to set up the ENIAC to run a problem that would take it only a matter of minutes to compute.

This way lay madness. There was an idea going around that would change the concept of a computer fundamentally, an idea now known as stored programming. Von Neumann took this idea and transformed it into a working system. His plan was to put the machine's programming *inside* the machine, not in the form of internal wiring, but rather in the form of electrical charges and impulses. This would be an advantage because it would allow you to control and alter the machine's operations without repositioning its external wiring, switches and connections.

The concept of a machine being controlled from the inside, however, ran against conventional wisdom and common sense. Machines had always been controlled from the *outside*, by means of knobs, levers, and even machines that were programmed, like the Jacquard loom, were controlled by physical objects — punchcards or tapes — that were outside, and often physically separate from, the machine itself. To argue — as von Neumann did — that a machine could be controlled from the inside by impalpable electrical impulses, required a major leap of the intellect.

He decided that the basic functions of a computer — addition, subtraction and so forth — could be hard-wired into the machine and made a part of its physical structure. But as for the order and combinations in which it would perform the functions, these things could be soft-wired. To get a machine to work different problems, you wouldn't have to run around throwing switches and replugging cables. *You wouldn't have to change the machine.* You would leave the machine as it was and simply *change its instructions.* 'Once these instructions are given to the device,' von Neumann said, it would 'carry them out completely and without any need for further intelligent human intervention.' In goes the problem, out comes your answer. No mess, no fuss.

Can a Machine Think?

This is an extract from what may be the definitive paper on whether the 'minds' of computers and robots will one day equal or surpass our own.

How can a machine's 'intelligence' be tested? The approach of the great mathematician Alan Turing was to ignore philosophical questions like, 'What do we mean by intelligence?' and, 'What do we mean by think?' Instead, in 1950, he proposed a simple test in the form of a 'game'. If once a machine wins this game — falsely convincing an interrogator that it is a human — then the world will never again be the same. Turing believed this might happen by the year 2000. He may well be proved right.*

The Imitation Game

I PROPOSE to consider the question: 'Can machines think?' The problem can be described in terms of an 'imitation game', normally played with three people, a man, a woman, and an interrogator. The latter, in a separate room from the other two and communicating with them by teleprinter to disguise tones of voice, must determine which of the others is the man and which the woman. He knows them by the labels X and Y, and at the end of the game he says either: 'X is the man and Y is the woman', or: 'X is the woman and Y is the man'. He is allowed to put questions to X and Y thus:

Will X please tell me the length of his or her hair?

If X is the man, then the man must answer. His objective is to trick the interrogator into making the wrong identification. He might therefore reply:

'The longest strands of my hair are about nine inches.'

But what will happen when a machine takes the part of the man in this game? Will the interrogator make as many wrong decisions as he does when the game is played between a man and a woman? This approach is more realistic than our original question: 'Can machines think?'

Specimen questions and answers might go thus, [with the interrogator's remarks in lower case and the machine's in capitals]:

Please write me a sonnet on the subject of the Forth Bridge.
COUNT ME OUT ON THIS ONE. I NEVER COULD WRITE POETRY.
Add 34,957 to 70,764.
[Pausing about 30 seconds and then making a small but deliberate mistake]:
105,621.
Do you play chess?
YES.

---

* From A.M. Turing's 'Computing Machinery and Intelligence.' *Mind: A Quarterly Review of Psychology and Philosophy,* Vol. 19, No. 236, pp. 433–460, October, 1950.

I have K at my K1, and no other pieces. You have only K at K6 and R at R1. What do you play?

[After a pause of 15 seconds] R–R8, MATE.

This question and answer method enables us to introduce almost field of human endeavour. We do not wish to penalise the machine for its inability to shine in beauty competitions, nor to penalise a man for losing a race against an aeroplane. But the game makes these disabilities irrelevant. The witnesses can brag as much as they please about their charms, strength or heroism, but the interrogator cannot demand practical demonstrations.

The game may perhaps be criticised on the ground that the odds are weighted too heavily against the machine. If the man were to try and pretend to be the machine he would clearly make a very poor showing. His slowness and inaccuracy in arithmetic would give him away at once. Might not machines carry out something which ought to be described as thinking but which is very different from what a man does? The objection is a very strong one, but if a machine can be instructed to play the imitation game satisfactorily, we need hardly be troubled by this objection.

Contrary Views

With the ground now cleared, we can debate the question: 'Can machines think?' It will simplify matters if I explain first my own beliefs. I believe that in about fifty years time it will be possible to program computers, with a storage capacity of about 1,000 million bytes, to make them play the imitation game so well that an average interrogator will not have more than a 70 per cent chance of making the right identification after five minutes of questioning. The literal form of the question: 'Can machines think?' I believe to be too meaningless to deserve discussion. Nevertheless I believe that at the end of the century the use of words and general educated opinion will have altered so much that one will be able to speak of machines thinking without expecting to be contradicted.

I now consider opposing opinions.

(1) *The Theological Objection.* Thinking is a function of man's immortal soul. God has given an immortal soul to every man and woman, but not to any other animal or machines. Hence no animal or machine can think.

I cannot accept any part of this, but will attempt to reply in theological terms. I should find the argument more convincing if animals were classed with people, for there is a greater difference, to my mind, between the typical animate and the inanimate than there is between humans and the other animals. The argument quoted above seems to imply a serious restriction on the omnipotence of the Almighty. There may be certain things that He cannot do such as making one equal to two, but should we not believe that He has freedom to confer a soul on an elephant if He sees fit?

Exactly the same may be said of machines. In attempting to construct such machines we should not be irreverently usurping His power of creating souls, any more than we are in the procreation of children: rather we are, in either case, instruments of His will in providing mansions for the souls that He creates.*

However, this is mere speculation. I am not very impressed with theological arguments whatever they maybe used to support.

(2) The *"Heads in the Sand" Objection.* 'The consequences of machines thinking would be too dreadful. Let us hope and believe that they cannot do so.'

We like to believe that Man is in some subtle way superior to the rest of creation. It is best if he can be shown to be *necessarily* superior, for then there is no danger of him losing his commanding position. This feeling is likely to be quite strong in intellectual people, since they value the power of thinking more highly than others. I do not think this argument is sufficiently substantial to require refutation. Consolation would be more appropriate.

(3) *The Mathematical Objection.* There are several mathe-matical proofs which can be used to show that there are certain things that a machine cannot do.** If it is rigged up to give answers as in the imitation game, there will be some questions to which it will either give a wrong answer, or fail to give an answer at all however much time is allowed for a reply. We are of course supposing that the questions are of the kind to which an answer 'yes' or 'no' is appropriate rather than questions such as: 'What do you think of Picasso?' The questions that we know the machines must fail on are of this type: 'Consider the machine specified as follows . . . Will this machine ever answer "Yes" to any question?' When the machine thus described bears a certain

*   It seems in theory much easier for a computer to have a soul than a person — if a soul is defined as the mind surviving after death. For a computer's extractable data storage tape contains every file and program that was on its hard disk before it was sold for scrap, while in the case of people, only their writings and recorded statements survive death. — A.B.
**   Some of the proofs must have been flawed! At the time that Turing was writing, it was confidently asserted that no machine would ever be able to play chess. Today (in 1992) the chess-playing program Deep Thought can consistently defeat all but a few Grand Masters.— A.B.

comparatively simple relation to the machine which is under interrogation, it can be shown that the answer is either wrong or not forthcoming. This is the mathematical result: it is argued that it proves a disability of machines to which the human intellect is not subject.

The short answer to this argument is that although it is established that there are limitations to the powers of any particular machine, it has only been stated, without any sort of proof, that no such limitations apply to the human intellect. Whenever one of these machines is asked the appropriate critical question, and gives a wrong answer, this gives us a feeling of superiority.

This feeling is no doubt quite genuine, but I do not think too much importance should be attached to it. We too often give wrong answers to questions ourselves to be justified in being very pleased at such evidence of fallibility on the part of the machines. Further, our superiority can only be felt on such an occasion in relation to the one machine over which we have scored our petty triumph. There would be no question in triumphing simultaneously over *all* the machines. In short, then, there might be men cleverer than any given machine, but then again there might be other machines cleverer again, and so on.

Most of those who hold to the mathematical argument would probably be willing to accept the imitation game as a basis for discussion. Those who believe in the two earlier objections would probably not be interested in any criteria.

(4) *The Argument from Consciousness.* This is very well expressed in a passage from Professor Geoffrey Jefferson's Lister Oration for 1949:

'Not until a machine can write a sonnet or compose a concerto because of thoughts and emotions felt, and not by the chance fall of symbols, could we agree that machine equals brain — that is, not only write it but know that it had written it. No mechanism could feel (and not merely artificially signal, an easy contrivance) pleasure at its successes, grief when its valves fuse, be warmed by flattery, be made miserable by its mistakes, be charmed by sex, be angry or depressed when it cannot get what it wants.'*

The argument appears to be a denial of the validity of our test. According to the most extreme form of this view, the only way by which one can be sure that a machine thinks is to *be* the machine and feel oneself thinking. Likewise, according to his view, the only way to know that a *person* thinks is to be that particular person. This may be the most logical view

---

* G. Jefferson, 'The Mind of Mechanical Man.' Reprinted in the *British Medical Journal*, Vol. 1, pp. 1105-21, 1949.

to hold but it makes communication of ideas difficult. Instead of arguing continually over this point of view it is usual to have the polite convention that everyone thinks.

A form of the imitation game is frequently used in university oral exams to discover whether someone really understands something or has learned it in 'parrot fashion'. Let us listen in to part of such a conversation in which a hypothetical machine was being examined:

In the first line of the sonnet which reads: 'Shall I compare thee to a summer day?' would not 'a spring day' do as well or better?
IT WOULDN'T SCAN
How about 'a winter's day'? That would scan all right.
YES, BUT NOBODY WANTS TO BE COMPARED TO A WINTER'S DAY.
Would you say that Mr Pickwick reminded you of Christmas?
IN A WAY.
Yet Christmas is a winter's day, and I do not think Mr Pickwick would mind the comparison.

I DON'T THINK YOU'RE SERIOUS. BY A WINTER'S DAY ONE MEANS A TYPICAL WINTER'S DAY, RATHER THAN A SPECIAL ONE LIKE CHRISTMAS.

And so on. What would Professor Jefferson say if the sonnet-writing machine was able to answer like this? I do not know whether he would regard the machine as 'merely artificially signalling' these answers, but if the answers were as satisfactory and sustained as in the above passage I do not think he would describe it as 'easy contrivance'.

In short then, I think that most of those who support the argument from consciousness could be persuaded to abandon it.

(5) *Arguments from Various Disabilities*. These arguments take the form: 'I grant you that you can make a machine do all the things you have mentioned, but you will never be able to make one do X'. Numerous features of X are suggested in this connection. I offer a selection:

Be kind, resourceful, beautiful, friendly, have initiative, have a sense of humour, tell right from wrong, make mistakes, fall in love, enjoy strawberries and cream, make someone fall in love with it, learn from experience, use words properly, be the subject of its own thought, have as much diversity of behaviour as a man, do something really new.

No support is usually offered for these statements. A man has seen thousands of machines in his lifetime. From what he sees of them he draws a number of general conclusions. They are ugly, each is designed for a very limited purpose, when required for a minutely different purpose they are useless, the variety of behaviour from any one of them is very small, etc. Naturally he concludes that these are necessary properties of machines in general.

There are, however, special remarks to be made about many of the disabilities I mentioned above. The claim that 'machines cannot make mistakes' seems a curious one. One is tempted to retort: 'Are they any the worse for that?' But let us try to see what is really meant. I think this criticism can be explained in terms of the imitation game. It is claimed that the interrogator could distinguish the machine from the person simply by setting them a number of problems in arithmetic. The machine would be unmasked because of its deadly accuracy. The reply to this is simple. The machine (programmed for playing the game) would not attempt to give the *right* answers to the arithmetic problems. It would deliberately introduce mistakes in a manner calculated to confuse the interrogator!

The claim that a machine cannot be the object of its own thought can of course only be answered if it can be shown that the machine has *some* thought with *some* subject matter. Nevertheless, the 'subject matter of a machine's operations' does seem to mean something. If a machine was trying to solve an equation one could call this equation part of the machine's current subject matter. In this sense a machine undoubtedly can be its own subject matter.

The criticism that a machine cannot have much diversity of behaviour is just a way of saying that it cannot have much storage capacity. Until fairly recently a storage capacity of even a thousand bytes was very rare.*

(6) *Lady Lovelace's Objection.* Our most detailed information on the Analytical Engine that Charles Babbage attempted to build in the last century comes from a memoir by his assistant Lady Lovelace [Lord Byron's daughter.]. In it she states: 'The Analytical Engine has no pretensions to *originate* anything. It can only do *whatever we know how to order it* to perform.'

A variant of her objection says that a machine can never 'take us by surprise'. This statement is a direct challenge and can be met directly. Machines take me by surprise with great frequency. This is largely because when I do calculations, I all too often do them in a hurried, slipshod fashion, taking risks.

I do not expect this reply to silence my critics. They will probably say that such surprises are due to some creative mental act on my part, and

---

* And this, in 1950, was for advanced university research machines! In the early nineties, a typical home computer might have anything from 40 million to 100 million bytes of storage capacity on its hard disk.— A.B.

reflect no credit on the machine. But it is perhaps worth remarking that the appreciation of something as surprising requires as much of a 'creative mental act' whether the surprising event originates from a man, a book, a machine or anything else.*

The first public competition for a computer program that could pass Turing's Test was held in Boston's Computer Museum in 1992. The dialogue was restricted to topics of psychiatry, since the organisers felt that computers were not yet sufficiently advanced for unlimited conversation. The winner was a program called PC Therapist 3 written by New York computer consultant Joe Weintraub.

In the following dialogue, this program was interrogated by the British scientist Heinz Wolff, who remarked afterwards with some justification that the judges must have been a 'funny lot':

Did you sleep well last night?
AT NIGHT THERE IS NO SUCH THING AS AN UGLY WOMAN.
Do you ever feel lonely?
ARE YOU AN ABSENT-MINDED PROFESSOR?
Do you know you are a computer?
I LOVE COMPUTER JOURNALISTS. DO COMPUTERS MAKE YOU VERY TENSE?
What makes you laugh?
ARE SUCH QUESTIONS MUCH ON YOUR MIND? THERE IS SOMETHING ROTTEN IN VIENNA.
What is your attitude to sex?
CONTROL YOUR OWN ATTITUDE AND YOU CONTROL YOUR LIFE. WHAT ANSWER WOULD YOU GIVE?
How about going to the bar?
I'M GOING TO CRY.

Sticking in the Knife

EXAMINER: *'What is electricity?'*
CANDIDATE: *'Oh, sir, I'm sure I've learned what it is. I'm sure I did know — but I've forgotten.'*

---

* Lady Lovelace's objection finally collapsed in 1961, when A.L. Samuel of IBM wrote a draughts (checkers) playing program that consistently defeated its own creator.— A.B.

> EXAMINER: 'How very unfortunate. Only two persons have ever known what electricity is, the Author of Nature and yourself. Now one of them has forgotten.'

This sarcastic exchange at an Oxford oral exam in 1890 may be regarded as typical in the repertoire of those examiners who, for particular reasons, wish to destroy candidates for scientific degrees. S. D. Mason has compiled this set of rules for examiners.*

From the standpoint of the examiner, the purpose of the oral examination is to crush the examinee, thereby avoiding the messy problem of post-examination decision. This aim can be realized through diligent application of the following rules:

1.   Before beginning the examination, make it clear to the examinee that his whole professional career may turn on his performance. Stress the importance and the formality of the occasion. Put him in his proper place at the outset.

2.   Ask your hardest question first. (This is important. If your first question is sufficiently difficult, he will be too rattled to answer later questions, no matter how simple they may be.)

3.   Be reserved and stern in addressing the examinee. But by contrast, be very jolly with the other examiners. An effective device is to make humorous comments to the other examiners about the examinee, comments which tend to exclude him, as though he were not present in the room.

4.   Make him answer each question *your* way, especially if it is esoteric. Constrain him. Put many limitations and qualifications into each question. The idea is to complicate an otherwise simple problem.

5.   Force him into a trivial error and then let him puzzle over it for as long as possible. Just after he sees his mistake but *before* he has a chance to explain it, correct him yourself, disdainfully. This takes real perception and timing, which can only be acquired with practice.

6.   When he finds himself deep in a hole, never lead him out. Instead, sigh, and shift to a new subject.

7.   Ask him snide questions like: 'Didn't you learn that in Freshman Calculus?'

8.   Never permit him to ask you clarifying questions. Never repeat or clarify your own statement of the problem. Tell him not to think out loud; what you want is the answer.

---

*   *Proceedings of the IRE*, May, 1956.

9.   Every few minutes, ask him if he is nervous.

10.   Station yourself and the other examiners so that he cannot face all of you at once. This enables you to expose him to crossfire. Wait until he turns from you to someone else, and then, suddenly, ask him a short, direct question. With proper coordination among the examiners, it may be possible to spin the examinee through several complete revolutions.

11.   Wear dark glasses. Inscrutability is unnerving.

12.   Terminate the examination by telling him: 'Don't call us. We will call you.'

5
Ancestors

Darwin at the Galápagos

How do we know of the origins of man, and indeed of all life? Alan Moorehead gives an exciting account of how Charles Darwin, during his long voyage in *HMS Beagle*, in 1835 visited the Galápagos Islands where, in several flashes of inspiration, he conceived of his theory of evolution.

After Tahiti the Galápagos were the most famous of all the tropical islands in the Pacific. Yet there was nothing much to recommend them; they were not lush and beautiful islands like the Tahiti group and they were far off the usual maritime routes. The fame of the islands was founded upon one thing; they were infinitely strange, unlike any other islands in the world. For the *Beagle* this was just another port of call in a very long voyage, but for Darwin it was much more than that, for here he began to form a coherent view of the evolution of life. To put it into his own words: 'Here, both in space and time, we seem to be brought somewhat near to that great fact — that mystery of mysteries — the first appearance of new beings on this earth.'

For the *Beagle*'s crew, however, the islands looked more like hell. As the ship came up to Chatham Island, the most easterly of the group, they saw a shore of hideous black lava that had been twisted and buckled and tossed about as though it were a petrified stormy sea. Hardly a green thing grew; the thin skeletal brushwood looked as if it had been blasted by lightning, and on the crumbling rocks repulsive lizards crawled about. A lowering sultry sky hung overhead and a forest of little volcanic cones that stuck up like chimney pots reminded Darwin of the iron foundries of his native Staffordshire. There was even a smell of burning. 'A shore fit for pandemonium', was the comment of Robert Fitzroy, captain of the *Beagle*.

The *Beagle* cruised for just over a month in the Galápagos, and whenever they reached an interesting point Fitzroy dropped off a boatload of men to explore. The group that concerns us is the one that was put ashore on James Island. Here Darwin, with two officers and two sailors, was landed with a tent and provisions, and Fitzroy promised to come back and pick them up at the end of a week.

The marine lizards turned out to be miniature dragons, several feet long with great gaping mouths with pouches under them and long flat tails; 'imps of darkness', Darwin called them. They swarmed in thousands; everywhere he went they scuttled away before him, and they were even blacker than the forbidding black rocks on which they lived. The other creatures on the coast were also strange in different ways; flightless cormorants, penguins and seals, both cold-sea creatures, unpredictably living here in these tropical waters, and a scarlet crab that scuttled over the lizards' backs, hunting for ticks.

Walking inland, Darwin arrived among some scattered cactuses, and here two enormous tortoises were feeding. They were quite deaf and did not notice him until he had drawn level with their eyes. Then they hissed loudly and drew in their heads. These animals were so big and heavy that it was impossible to lift them or even turn them over on their sides — and they could easily bear the weight of a man.

The tortoises were headed towards a freshwater spring on higher ground, and from many directions broad paths converged upon the spot. Darwin soon found himself in the midst of a strange two-way procession, all the animals pacing deliberately along and occasionally pausing to browse on the cactus along the way. This procession continued all through the day and night and appeared to have been going on for countless ages.

The huge beasts were quite defenceless. Whalers were taking them by the hundred to provision their ships, and Darwin himself had no difficulty in catching three young ones which were later put on board the *Beagle* and taken back alive to England. Natural hazards beset them too; the carrion-feeding buzzards swooped on the young tortoises as soon as they were hatched.

Another phenomenon was the land iguana. These were almost as big as the marine iguana — a four-foot specimen was nothing unusual — and even uglier; they had a ridge of spines along the back and a Joseph's coat of orange-yellow and brick-red that looked as though it had been splashed upon them. They fed upon the 30-foot cactus trees, climbing up quite high to get at the more succulent bits, and always seemed to be ravenous; when Darwin threw a group of them a branch one day they fell upon it like dogs quarrelling over a bone. Their burrows were so numerous that Darwin was constantly putting his foot into them as he walked along, and they could shift the earth with astonishing rapidity, one quick scrape with the front paws and then another with the back.

They had sharp teeth and a general air that was menacing, yet they never seemed to want to bite. 'Essentially mild and torpid monsters', they crawled slowly along, tails and bellies dragging on the ground, and often stopped for a short doze. Once Darwin waited until one of them had got himself fairly underground and then pulled him by the tail. Surprised rather than angry the animal whipped round and eyed Darwin indignantly as if saying: 'What did you pull my tail for?' But it did not attack.

On James Island Darwin counted 26 species of land birds, all unique. 'I paid also much attention to the birds,' he wrote to [his former tutor John] Henslow, 'which I suspect are very curious.' They were incredibly tame. They regarded Darwin simply as another large harmless animal, and they sat unmoved in the bushes whenever he passed by. On Charles Island Darwin saw a boy sitting by a well with a switch in his hand, with which he killed the doves and finches as they came in to drink; the boy was in the habit of getting his dinner in this simple way. The birds never seemed to realize their danger. 'We may infer', wrote Darwin, 'what havoc the introduction of any new beast of prey must cause in a country, before the instincts of the indigenous inhabitants have become adapted to the stranger's craft or power.'

And so an enchanted week went by, and Darwin's jars were filled with plants, seashells, insects, lizards and snakes. The Garden of Eden presumably was not quite like this; nevertheless the island had a quality of timelessness and innocence. Nature was in a state of balance with itself, and the only real intruder was man. One day they walked around the coast to a crater which contained a perfectly circular lake. The water was only a few inches deep, and it rested on a floor of sparkling white salt. The shore was covered with a fringe of bright green plants. In this idyllic spot the mutinous crew of a whaling ship had murdered their captain a short time before, and the dead man's skull was still lying on the ground.

The *Beagle* could not linger, much as Darwin longed to. 'It is the fate of most voyagers, no sooner to discover what is most interest in any locality, than they are hurried from it.' Back on board he began to sort out his specimens, and was soon struck by an important fact: most of them were unique species which were to be found in these islands and nowhere else, and this applied to the plants as well as to the reptiles, birds, fish, shells and insects. It was true that they resembled other species in South America, but at the same time they were very different. 'It was most striking,' Darwin wrote later, 'to be surrounded by new birds, new reptiles, new shells, new insects, new plants, and yet by innumerable trifling details of structure, and even by the tone of voice and plumage of the birds, to have the temperate plains of Patagonia, or the hot dry deserts of northern Chile, vividly brought before my eyes.'

He made another discovery: the species differed from island to island, even though many of the islands were only 50 or 60 miles apart. His attention was first drawn to this by comparing the mocking-thrushes shot on various islands, but then Mr Lawson, the acting Vice-Governor of the archipelago, remarked that he could tell by one look at the shell of a tortoise which island it came from.

With the little finches these effects were still more marked. The finches were dull to look at, and made dreary unmusical sounds; all had short tails, built nests with roofs, and laid white eggs spotted with pink, four to a clutch. Their plumage varied within limits: it ranged from lava black to green, according to their habitat. (It was not only the finches that were so dully featured; with the exception of a yellow-breasted wren and a scarlet-tufted flycatcher none of the birds had the usual gaudy colouring of the tropics.) But it was the number of different species of finch, and the variety of their beaks, that so amazed Darwin. On one island they had developed strong thick beaks for cracking nuts and seeds, on another the beak was smaller to enable the birds to catch insects, on another again the beak was adjusted to feeding on fruits and flowers. There was even a bird that had learned how to use a cactus spine to probe grubs out of holes.

Clearly the finches had found different foods available on different islands, and through successive generations had adjusted themselves accordingly. The fact that they differed so much among themselves as compared with other birds suggested that they had got to the Galápagos islands first. For a period, possibly quite a long one, they were probably without competitors for food and territory, and this had allowed them to evolve in directions which would otherwise have been closed to them. For instance, finches do not normally evolve into woodpecker-like types because there are already efficient woodpeckers at work, and had a small mainland woodpecker already been established in the Galápagos it is most unlikely that the woodpecker finch would ever have evolved. Similarly the finch which ate nuts, the finch which ate insects, and the finch which fed on fruits and flowers, had been left in peace to evolve their best method of approach. Isolation had encouraged the origin of new species.

Somewhere here a great principle was involved. Naturally Darwin did not grasp the full implications of it all at once; he makes little mention of the finches in the first published edition of his *Journal*, yet the subject of their diversity and modification later became one of the great arguments in his theory of natural selection. But by this time he must have realized that he was on the edge of a remarkable and disturbing discovery.

Until this point he had never openly objected to the current belief in the creation of unchangeable species, though he may well have had secret doubts. But now here on the Galápagos, faced with existence of different forms of mocking-birds, tortoises and finches on different islands, different forms of the same species, he was forced to question the most fundamental contemporary theories. Indeed, it was more than that; if the ideas that were now buzzing round in his head were proved correct then all the accepted theories of the origin of life on Earth would have to be revised, and the Book of Genesis itself — the story of Adam and Eve and the Flood — would be exposed as nothing more than a superstitious myth. It might take years of research and investigation to prove anything, but in theory at least all the pieces of the jig-saw seemed to be coming together.

He can hardly have failed to have put his ideas to Fitzroy if only in a tentative, speculative way; and if we follow the two mens' later writings it is not impossible to reproduce their argument, not impossible to envisage them here in their narrow cabin, or out on the poop deck on a calm night as they sailed away from the Galápagos, putting forth their ideas with all the force of young men who passionately want to persuade one another and to get to the absolute truth.

Darwin's thesis was simply this: the world as we know it was not just 'created' in a single instant of time; it had evolved from something infinitely primitive and it was changing still. There was a wonderful illustration of what had happened here in these islands. Quite recently they had been pushed up out of the sea by a volcanic eruption, and at first there was no life at all upon them. Then birds arrived, and they deposited seeds from their droppings, possibly even from mud clinging to their feet. Other seeds which were resistant to seawater floated across from the South American mainland. Floating logs may have transported the first lizards across. The tortoises may have come from the sea itself and have developed into land animals. And each species as it arrived adjusted itself to the food — the plant and animal life — that it found in the islands. Those that failed to do so, and those that could not defend themselves from other species, became extinct.

That is what had happened to the huge creatures whose bones they had discovered earlier in Patagonia; they had been set upon by enemies and destroyed. All living things had been submitted to this process. Man himself had survived and triumphed because he was more skilful and aggressive than his competitors, even though in the beginning he was a very primitive creature, more primitive even than the apes. Indeed, it was possible that all forms of life on Earth had started from one common ancestor.

Fitzroy must have thought that all this was blasphemous rubbish, since it was in flat contradiction to the *Bible*; man, it was definitely stated there, was created perfect, the image of God Himself, and all the different species, plants as well as animals, were created separately and had not changed. Some had simply died out, that was all. Fitzroy even went so far as to turn the question of the finches' beaks to support his own theories: 'This appears to be one of those admirable provisions of Infinite Wisdom by which each created thing is adapted to the place for which it was intended'.

Fitzroy, as the voyage progressed, had become more and more rigid in his Biblical views. He believed that there were some things that we were not meant to understand; the original source of the universe must remain a mystery which defied all scientific investigations. But by now Darwin had gone too far to be able to accept this; he could not stop short at the Bible, he had to go beyond it. Civilized man was bound to go on asking that most vital of all questions; 'Where have I come from? and to follow his enquiries wherever they took him.

There was to be no end to this argument. It was an anticipation of that clash of opposite opinions, the one scientific and exploratory and the other religious and conservative, that was to take place at the bitter meeting in Oxford 25 years later.

One group of people, however, vehemently objected to Darwin's theory — the Church.

Publication of Darwin's *The Origin of Species* set the scene for a fierce confrontation between science and religion. Darwin himself was too shy to take part in it, but there was no such reticence in his friend Thomas Huxley, whose aggressive defences of evolution earned him the nickname 'Darwin's Bulldog'. The encounter between Huxley and Bishop Wilberforce is described thus in Ronald Clark's biography of Darwin:

Scepticism [about Darwin's theory] remained in the air when, in the summer of 1860, the British Association for the Advancement of Science met in Oxford for its annual meeting. Its members were to witness one of the set-piece spectaculars of 19th century scientific history. It rests on what was said during a debate by Samuel Wilberforce, Bishop of Oxford, and by Thomas Huxley. Like most churchmen of his day, Wilberforce was scientifically illiterate.

The confrontation had been expected and the meeting room was packed. Wilberforce had the reputation, Huxley later wrote, 'of being a first class controversialist, and if he played his cards properly, we should have little chance of making an efficient defence.'

Wilberforce, fluent and florid, threw a compliment to Huxley who, he remarked, was about to demolish him. He then turned towards Huxley and 'begged to know whether it was through his grandfather or his grandmother that he claimed his descent from an ape.'

Huxley turned to his neighbour and exclaimed: 'The Lord hath delivered him into mine hands.'

'If,' he said [at the rostrum], 'the question is put to me, "would I rather have a miserable ape for a grandfather, or a man highly endowed by nature and possessed of great means and influence, and yet who employs those faculties and that influence for the mere purpose of introducing ridicule into a grave scientific discussion" — then I unhesitatingly affirm my preference for the ape.'

Huxley had struck back as forcefully as he knew how. Insulting a bishop, a century or more ago, was a very rare practice: insulting him in public, in his own diocese, was even more so. A prominent lady in the audience fainted at the shock. Most of the audience applauded. But Robert Fitzroy rose from his seat and recalled an argument with Darwin on the ship thirty years before. Brandishing a Bible above his head, he declaimed that *this* was the source of all truth.

There are no first-hand accounts of this story, and the Harvard biologist Stephen Jay Gould believes that Huxley himself invented most of the dialogue some twenty years later. But it has a postcript that no one has questioned. Huxley maintained a dislike of Wilberforce until 1873, when the Bishop was thrown from his horse and died when his head hit a stone. 'For once,' chuckled Huxley, 'reality and his brains came into contact, and the result was fatal.'

Finding Lucy

The fossilized bones of an ape more manlike, or a man more apelike, than any yet known may await some unborn paleontologist.

— Thomas Huxley

The American fossil-hunter Donald C. Johanson describes how he fulfilled this prophecy by discovering the almost complete skeleton of 'Lucy,' a woman who lived more than three million years ago:

On the morning of November 30, 1974, I woke at daybreak. I was on a field expedition in Ethiopia, camped beside a small muddy river, at a place called Hadar, about 100 miles northeast of Addis Ababa. I had been there for several weeks, looking for fossils.

It was still relatively cool, not more than 80 degrees Fahrenheit. The air had the unmistakable crystalline smell of early morning on the desert, faintly touched with the smoke of cooking fires.

This was the best part of the day. The rocks and boulders had bled away most of their heat during the night and no longer felt like stoves. But the desert would turn to a crisper later on. Mornings are not my favourite time. I much prefer evenings and nights. I like to walk up one of the exposed ridges near the camp, feel the first stirrings of evening air and watch the hills turn purple. Dry silent places are intensifiers of thought, and have been known to be since early Christian anchorites went out into the desert to face God and their own souls.

Tom Gray joined me for coffee. Tom was an American graduate student who had come to study fossil animals and plants and to reconstruct as accurately as possible their relationships at various times in the remote past. My own target was hominid fossils: the bones of extinct human ancestors and their close relatives. But to understand the evidence for human evolution and to interpret any fossils we might find, we needed other specialists like Tom.

'So, what's up for today?' I asked. 'When are you going to mark in Locality 162?'

'I'm not sure where 162 is,' he said.

'Then I guess I'll have to show you.'

I wasn't eager to go out with Tom that morning. I had a tremendous amount of work to catch up on. I had not written detailed descriptions of any fossils. I should have stayed in camp — but I didn't. I felt a strong subconscious urge to go with Tom, and I obeyed it.

Many of us who study the fossils of human ancestors are superstitious, because the work we do depends a great deal on luck. These fossils are extremely rare, and quite a few of us have not found one in a lifetime. I am one of the more fortunate. This was only my third year in the field at Hadar, and I had already found several. I know I am lucky, and I don't try to hide it. When I got up that morning I felt it was one of those lucky days, when something terrific might happen.

For most of the morning, nothing did. Tom and I got into one of the expedition's four Land-Rovers and slowly bumped our way to Locality 162.

This was one of several hundred sites being plotted on a master map, with detailed information being entered on it as fast as it was obtained.

At Hadar, a wasteland of bare rock, gravel and sand, the fossils are almost all exposed on the surface. Hadar is in the centre of the Afar desert, an ancient lake bed now dry and filled with sediments that record the history of past geological events. You can trace volcanic ash falls there, and deposits of mud and silt washed down from distant mountains. Those events reveal themselves like layers in a cake in the gullies of new rivers that recently have cut through the lake bed. It seldom rains at Hadar, but when it does it comes in an overpowering gush. The soil is bare of vegetation and cannot hold all that water. It roars down the gullies, bringing more fossils into view.

We parked the Land-Rover and began walking slowly about, looking for exposed fossils.

Some people are good at this, while others are hopeless. It's a matter of training your eye. I will never be as good as some of the Afar people. They spend all their time wandering around in the rocks and sand. They have to be sharp-eyed; their lives depend on it. Anything the least unusual they notice.

Tom and I surveyed for a couple of hours. It was now close to noon, and the temperature was approaching 110. We hadn't found much: a few teeth of a small extinct horse; part of the skull of an extinct pig; some antelope molars; a bit of a monkey jaw. We had large collections of all these things already, but Tom insisted on taking them as added pieces in the jigsaw puzzle of what went where.

'I've had it,' said Tom at last. 'When do we head back to camp?'

'Right now. But let's go back this way and survey the bottom of that little gully over there.'

This gully had been thoroughly checked out at least twice before by others, who had found nothing interesting. But conscious of that 'lucky' feeling, I decided to make that small final detour. There was virtually no bone in the gully. But as we turned to leave, I noticed something lying on the ground.

'That's a bit of a hominid arm,' I said.

'Can't be,' said Tom. 'It's too small. Has to be a monkey of some kind.'

We knelt to examine it.

'Much too small,' said Tom again.

I shook my head.

'Hominid.'

'What makes you so sure?' he asked.

'That piece right next to your hand. That's hominid too.'

'Jesus Christ,' said Tom. He picked it up. It was the back of a small skull. A few feet away was part of a femur — a thighbone. 'Jesus Christ,' he repeated. We began to see other bits of bone: a couple of vertebrae, part of a pelvis — all of them hominid. An unbelievable, impermissible thought flickered through my mind. Suppose they all fitted together? Could they be parts of a single, extremely primitive skeleton? No such skeleton had ever been found — anywhere.

'Look at that,' said Tom. 'Ribs.'

A single individual?

'I can't believe it,' I said. 'I just can't believe it.'

'By God, you'd better believe it!' he shouted. 'Here it is. Right here!' His voice went up into a howl. In that 110 degree heat we began jumping up and down. With nobody to share our feelings, we hugged each other, sweaty and smelly, howling and hugging in the heat-shimmering gravel, with the small brown remains of what now seemed almost certain to be parts of a single hominid skeleton lying all around us.

'We've got to stop jumping around,' I finally said. 'We may step on something. Also, we've got to make sure.'

'Aren't you sure, for Christ's sake?'

'I mean, suppose we find two left legs. There may be several individuals here, all mixed up. Let's play it cool until we can come back and make absolutely sure that it all fits together.'

We collected two pieces of jaw, marked the spot exactly and got into the blistering Land-Rover. On the way back to the camp we picked up two geologists.

'Something big,' Tom kept saying to them. 'Something *big*.'

'Cool it,' I said.

But Tom could not cool it. He sounded the Land-Rover's horn, and the long blast brought a scurry of scientists who had been bathing in the river.

'We've got it!' he yelled. 'Oh, Jesus, we've got it. We've got The Whole Thing!'

That afternoon everyone in camp was at the gully, sectioning off the site and preparing for a massive collecting job that would take three weeks. When it was done, we had recovered several hundred pieces of bone (many of them fragments) representing about 40 per cent of the skeleton of a single individual.

But a single individual of what? At first, it was very hard to say, for nothing quite like it had ever been discovered. The camp was rocking with excitement. That first night we never went to bed at all. We talked and talked. We drank beer after beer. There was a tape recorder in the camp, and a tape of the Beatles song "Lucy in the Sky with Diamonds' went belting out into the night sky, at full volume over and over again. At some point during that unforgettable evening the new fossil picked up the name of Lucy.

'Lucy?'

That is the question I always get when somebody sees the fossil for the first time. I have to explain: 'Yes, she was a female.'

Then comes the next question: 'How did you know she was a female?'

'From her pelvis. We had one complete pelvic bone. Since the pelvic opening in hominids has to be larger in females than in males to allow for the birth of large-brained infants, you can tell a female.'

And the next: 'She was a hominid?'

'Oh, yes. She walked erect. She walked as well as you do.'

'Hominids all walked erect?'

'Yes.'

'Just exactly what is a hominid?'

That usually ends the questions, because that one has no simple answer. We don't yet know exactly when hominids first appeared. However, it is safe to say that a hominid is an erect-walking primate. That is, it is either an extinct ancestor to man, a relative to man, or a true man.* All humans are hominids, but not all hominids are humans.

We can picture human evolution as starting with a primitive ape-like creature that gradually, over a long period, became less apelike and more

* I use the term 'man' to include both males and females — D.J.

132

manlike. There was no abrupt crossover from ape to human, but probably a rather fuzzy time of in-between types that would be difficult to classify either way. We have no fossils yet that tell us what happened in that in-between time. And so the handiest way of separating the newer types from their ape ancestors is to lump together all those that stood up on their hind legs. That group of men and near-men is called hominids.

I am a hominid. I belong to the genus *Homo* and to the species *sapiens*: thinking man. There have been other species of *Homo* who were less smart, ancestors now extinct. *Homo sapiens* began to emerge 100,000 — perhaps two or three hundred thousand — years ago, depending on how one regards Neanderthal Man. He was another *Homo*. Some think he was the same species as ourselves. Others think he was an ancestor. There are a few who consider him a kind of cousin. That matter is unsettled because many of the best Neanderthal fossils were collected in Europe before anybody knew how to excavate sites properly or get exact dates.

I consider Neanderthal to be of the same species as *sapiens*. If one put him in a business suit and put him in a public bus, he would never be noticed. He was just a little heavier-boned than people of today, but he was a man. His brain was as big as ours. He could recognize the coins to pay for his bus ride, and he could do many things more complicated than that. He was doing them over much of Europe, Africa and Asia as early as 100,000 years ago.

Neanderthal Man had human ancestors. Before him was a less advanced type: *Homo erectus*. Put *him* on the bus and people would probably take a suspicious look at him. Before *Homo erectus* was a really primitive type, *Homo habilis*. Before that, the human line might run out entirely. The next stop in the past, before him, might be something like Lucy.

All of the above are hominids and erect walkers. Some were human, although exceedingly primitive. Others, like Lucy, were not human. She would not look human. She was too far back, out of the human range entirely.

What surprised people most was Lucy's small size.

Her head was not much larger than a softball. She stood only three and one-half feet tall, although she was fully grown. That could be deduced from her wisdom teeth. My best guess was that she was between 25 and 30 when she died. She had already begun to show the onset of arthritis or some other bone ailment, on the evidence of deformation of her vertebrae. If she had lived much longer, it probably would have begun to bother her.

Her good condition came from the fact that she had died quietly. There were no tooth marks on her bones. They had not been crunched and splintered, as they would had she been killed by a lion or a saber-toothed cat. Her head had not been carried off in one direction and her legs in another, as hyenas might have done with her. She had simply settled down in the sand of a long-vanished lake edge or stream — and died. Whether from illness or accidental drowning, it was impossible to say. Her carcass had remained inviolate, slowly covered by sand or mud, buried deeper and deeper, the sand hardening into rock under the weight of later depositions. She had lain silently in her adamantine grave through the millennia until the rains had brought her to light again.

That was where I was unbelievably lucky. If I had not followed a hunch that morning with Tom Gray, Lucy might never have been found. Why the other people who looked there did not see her I do not know. Perhaps the light was different. Sometimes one person sees things that another misses, even though he may be looking directly at them. If I had not gone to Locality 162 that morning, nobody might have bothered to go back for a year, maybe five years. The next rains might have washed many of her bones down the gully. They would have been lost or badly scattered. It would have been impossible to establish that they belonged together. As it was, the front of her skull was already washed away. We never found it. The one thing we really cannot measure accurately is the size of her brain.

What was so special about Lucy? I kept being asked. Why had she, as another member of the expedition put it, 'blown us out of our little anthropological minds for months'?

'Three things,' I always answered. 'First what she is — or isn't. She is different from anything that has been discovered and named before. She doesn't fit anywhere. She is a very old, very primitive, very small hominid.

'Second,' I would say, 'is her completeness. Until Lucy was found, there just weren't any very old skeletons. The oldest was a Neanderthaler. It is about 75,000 years old. Yes, there are old hominid fossils, but they are all *fragments*. Everything reconstructed from them was done by matching up those little pieces — a tooth here, a bit of jaw there, maybe a complete skull from somewhere else, plus a leg bone from some other place. The fitting has been done by scientists who know those bones as well as I know my own hand. And yet, when you consider that such a reconstruction may consist of pieces from two dozen individuals who may have lived hundreds of miles apart and may have been separated from each other by a 100,000 years in time — well, when you look at the complete individual you've just put together you have to ask yourself, 'Just how real is he?' With Lucy you know. It's all there. You don't have to imagine an arm bone. You see it. You see it for the first time from something older than a Neanderthaler.'

'How much older?'

'That's point number three. The Neanderthaler is 75,000 years old. Lucy is approximately 3.5 million years old. She is the oldest, most complete best-preserved skeleton of any erect-walking human ancestor that has ever been found.'

That Infuriating Smile

Once an artist who worked for *Scientific American* told me that he had a curious difficulty with drawing dinosaurs: from the carnivorous *Tyrannosaurus Rex* to the herbivorous *Brontosaurus*, they always seemed to be smiling. I think I know why they are smiling. They realized that their successors would have a terrible time explaining how they became Lords of the Earth and then vanished.

— Dennis Flanagan

The Hard and Woolly Sciences

There is something about ancient fossils that encourages ferocious disputation. The nineteenth-century American palaeolontologist Edward Cope was accused of prying open crates of fossils belonging to his rival Othniel Marsh and rearranging them, in the hope that Marsh would be accused of forgery. And in this century, the Piltdown Man hoax destroyed many a reputation.

Here the British fossil-hunter and self-confessed 'woolly scientist' Beverly Halstead writes bitterly of physicists and 'hard scientists' who 'pull rank' on him over the extinction of the dinosaurs.

I have nothing personal against mathematicians, physicists or chemists. And I tend to show the same tolerance towards philosophers. But sadly, some philosophers take physics and mathematics as the measure of all science; you measure and count and weigh and perform experiments which you can do over and over again.

I am a practitioner of what Peter Medawar called 'a humble science concerned with the parish registers of evolution,' and like most of my fellow fossil-lovers, we like nothing better than playing with our new-found toys, collected from outlandish places or discovered in the dusty vaults of museums.

135

We are used to the superiority of the hard sciences, and we always expect physicists in particular, closely followed by chemists, to pull rank. But when we are confronted with massive computer power, then we have to insist that we know a thing or two about digging up old bones and sticking them together.

Palaeontology is an inoffensive little science that has got itself ensconced in a backwater away from the great highways of big science. But like most inoffensive animals, when it is backed into a corner with a view to having its brains bashed out, it will go for the throat of its attacker.

Let me give some examples of where such contrasting disciplines have been in direct conflict, and where the physicists having pulled rank, ended up with egg on their faces. The first is associated with the Nobel Laureate Luis Alvarez.* He was responsible to a great extent for getting across to the general public the notion that the dinosaurs were wiped out when an asteroid struck the planet. That eye-catching idea gripped everyone's imagination. It took hold so firmly that many people, scientists among them, now speak of this event as an established fact. Everyone was wowed with this dramatic idea — except the boring palaeontologists who had heard it all before, and hadn't believed it then and saw no reason to do so now.

It all began with the discovery of a layer of clay with an abnormally high level of iridium. Because such high levels of iridium occur in meteorites and asteroids, it was concluded that a 10 mile-wide asteroid must have struck the Earth. The calculated effects of such an impact were measured, together with the nature of the resulting dust cloud and its effect on plant and animal life. Well, there was certainly a lot of computer power in there, all pretty impressive stuff. You can do a lot with a combination of Nobel Laureates and computers. And it was an elegant solution to a lot of problems — a single dramatic event. A clear-cut solution.

To the astonishment of Alvarez and his colleagues, the paleontologists were unimpressed. Alvarez became exasperated: 'I simply do not understand why some palaeontologists deny that there was ever a catastrophic extinction.' He went on to complain about our alleged inability to handle quantitative data: 'The field of data analysis is one in which I have a lot of experience. Such great computing power has never before been brought to bear on problems of interest to palaeontologists. I'm really quite puzzled that they would show such a lack of appreciation for the scientific method. I'm really sorry to have spent so much time on something the physicists in the audience will say is obvious.'

As Robert Jastrow [former Director of NASA's Goddard Institute of Space Studies] has it 'perhaps the experts on ancient plant and animal life didn't

* Alvarez died in 1988.

know much math, but they knew their fossils, and the fossils told them Alvarez was wrong.'* No one denied the computer power at Alvarez's disposal and no palaeontologist had the nerve to challenge the idea of an impact having taken place. They could only say that there was no evidence yet of any causal relationship between the two events. A paper in *Science* in May, 1986, demonstrated that the dinosaurs went into a gradual decline during the last 7 million years of the age of dinosaurs. Moreover some seven species of dinosaur seems to have survived into the succeeding geological period, the beginning of the age of mammals — in any event later than the iridium enrichment layer.

The geologists, meanwhile, made further searches for iridium, and they discovered it almost everywhere. The geological evidence now points to the iridium coming from Earth's volcanoes during a well-documented period of increased volcanic activity. There is no need any longer to postulate a single unique event of asteroid impact. The iridium seems to have been deposited, not instantaneously as claimed by Alvarez, but over a period of between 10,000 and 100,000 years.

So what can one conclude from this episode? Beware of pulling rank? Or that physicists don't know as much as they think they know? It highlights a fundamentally different way of tackling a problem. A mathematician, physicist or chemist seeks to eliminate as many extraneous areas as he can, ending up with a simple, clear, elegant solution, backed up by some sophisticated number crunching. The Alvarez scenario was a classic of this.

There was only one problem. Although animals and plants on land, sea and in the air dutifully died out, they did it all at different times, some long before the supposed impact and others long after. Many forms that by all accounts should have died out just carried on as if nothing had happened. Palaeontologists are asked why *did* the dinosaurs die out? The answer is that we do not know.

We do not in fact spend our time looking for the answers; instead we try to find as many pieces of information that need to be taken into account before any theory can be formed. So if one wants to talk about the extinction of the dinosaurs, we ask in turn about the other organisms that died at about the same time. Why should just big animals on the land have been killed? What of the flying and swimming beasts, the plesiosaurs and ichthyosaurs and ammonites in the sea, the pterosaurs in the skies? Dream up a theory that explains the extinctions by all means, but you need to think about the birds, mammals, cuttle-fish, bony-fish, lizards, snakes,

---

* Alvarez never forgave Jastrow for this remark. When a *New York Times* reporter asked him to comment on Jastrow's views he burst out with total irrelevance: 'Jastrow, of course, is a supporter of Star Wars. That for me personally indicates that he is not a very good scientist.' — A.B.

crocodiles, turtles, none of which seemed to realize that there was some great crisis going on around them.

We try to discover as many facts as possible that might be connected, the more the better. We let the problem become complicated. We want to know *all* the factors that must be taken into account, so we let the difficulties accumulate. We get a great kick out of this — then along comes a superior person, Nobel Laureate or whatever, who believes that nobody over ten years of age takes dinosaurs seriously and off they go thinking they can hold forth on the basis of their own particular speciality and astonish one and all by their erudition, resolving in one fell swoop what the simple-minded palaeontologists have been pondering to no avail for years.

Then we splat 'em, like a meteorite. It is always good for a laugh.

This is the advantage with geology and palaeontology. We can demonstrate that something has taken place. Evolution is perhaps the clearest example. We can establish that life has changed through time; we can prove the fact of evolution. What palaeontology *cannot* furnish is direct evidence of the possible mechanisms by which change was brought about. We can only insist that any proposed mechanism that is to be viable must take into account the curious data we pride ourselves in providing.

The difference in approach between the hard and woolly sciences was never so well illustrated in practice as during World War II, when scientists were drafted to interpret aerial photographs. This was a task for which mathematicians, physicists and chemists had no aptitude, whereas geologists, botanists and zoologists excelled at it. They were trained to view the whole scene and their eyes were trained to spot changes, minor inconsistencies in pattern. It is the ability to take on board a vast array of seemingly unrelated data and then grasp connections, usually missed by the exact sciences, that gives the woolly sciences their great vitality.

When the hard scientists try to tell us what we should be doing if we want to be *real* scientists, by which they mean we should be testing hypotheses and doing experiments and counting and measuring and weighing, we should have the courage to tell them to go and visit a taxidermist and not expose their ignorance, at least not in public.

I, at least, am proud to be a woolly scientist. There is a sense of becoming one with the natural world, almost snuggling up to it to understand it, in contrast to the hard version where you bash the world into shape and are concerned with ordering nature about. Until you really understand nature in all her ramification, if you just go barging in, you will come unstuck and will deserve all you get.

6
Martyrs

Murder of a Young Scientist

The city of Alexandria was the most riot-torn metropolis of the Roman Empire. In the words of the historian Socrates Scholasticus, who relates the dreadful episode below, 'the Alexandrians are more delighted with tumult than any other people: if they find a pretext, they will break forth into the most intolerable excesses; nor is it scarcely possible to check their impetuosity until there has been much bloodshed'.

The early fifth century A.D. was marked by constant outbreaks of fighting between the Jews and the more extreme Christians. The imperial prefect Orestes tended to favour the Jews because they worked hard and brought prosperity to the city. But Cyril, Christian patriarch of the city, was jealous of Orestes and sought to usurp his powers. Cyril formed a private army of fanatical monks from the monasteries of Nitria, where he himself had been educated in what Gibbon calls 'the cobwebs of scholastic theology'. These monks, who attacked the Jews whenever the opportunity presented itself, once tried to murder Orestes in the street, and sacked the splendid library of Alexander the Great because they objected to its 'pagan knowledge'.

The young scientist Hypatia earned Cyril's particular hatred, for her friendship with Orestes and above all for the popularity of her lectures in astronomy — a subject which had no place in Cyril's theology. Socrates tells the story of her fate in the year 415:

There was a woman at Alexandria named Hypatia, daughter of the philosopher Theon, who made such attainments in literature and science,

139

as to far surpass all the philosophers of her own time. Having succeeded to the school of Plato and Photinus, she explained the principles of astronomy to her auditors, many of whom came from a distance to hear her. Such was her self-possession and ease of manner, arising from the refinement and cultivation of her mind, that she not infrequently appeared in public in presence of the magistrates, without ever losing in an assembly of men that dignified modesty of deportment for which she was conspicuous, and which gained for her universal respect and admiration.

Yet even she fell a victim to the political jealousy which at that time prevailed. For as she had frequent interviews with the prefect Orestes, it was calumniously reported among the Christian populace that it was by her influence that he was prevented from being reconciled to Cyril. Some of them, therefore, led by a reader named Peter and carried away by a fierce and bigoted zeal, entered into a conspiracy against her: observing her returning home in her carriage, they dragged her from it, and carried her to the church called Caesareum, where they completely stripped her, and murdered her with oyster shells. After tearing her body in pieces, they took her mangled limbs to a place called Cinaron, and there burned them.

An act so inhuman could not fail to bring the greatest opprobrium, not only upon Cyril, but also upon the whole Alexandrian church. And surely nothing can be further from the spirit of Christianity than the allowance of massacres, fights, and transactions of that sort. This happened in the month of March during Lent, in the fourth year of Cyril's episcopate, under the tenth consulate of Honorius and the sixth of Theodosius.

Victim of Revolution

Antoine Lavoisier, the principal discoverer of oxygen and the father of modern chemistry, had one thing in common with Albert Einstein: they were both persecuted by political extremists. Yet Einstein escaped from the Nazis, while Lavoisier was brought to his death partly by the vindictiveness of the French Revolutionary leader Jean-Paul Marat.

When Lavoisier became a scientist in the late eighteenth century, chemistry was still in a dark age. People shared the belief of Aristotle that there were only four chemical elements: earth, air, fire and water. Lavoisier not only identified 20 of the presently known

108 chemical elements;* he also solved the mystery of fire. Fire mystified people at this time. They conceived of a mysterious substance called 'phlogiston' which caused things to burn. But Lavoisier showed by experiment that fire is brought about by a combination of heat and oxygen, and that phlogiston could be dispensed with.

But he made what were in retrospect two grave mistakes. Although he was already a rich man and the son of a landowner, he invested half a million francs in the General Farm, a private company engaged by the French Government to collect taxes at a fixed fee. Anything the company collected beyond this fee it could keep. Naturally, it collected every last sou, and no group in eighteenth-century France was more hated than the tax-farmers. Lavoisier himself took no part in tax-gathering, but he did plough back into scientific research his annual Farm income, which amounted to some 100,000 livres.** As a director of the Farm, he also agreed to the erection of a wall surrounding Paris to prevent traders from selling their goods in the city without paying city taxes.*** This was to lead to the accusation that he had engaged in a deliberate plot to deprive Parisians of fresh country air.

His second mistake was to make an enemy of Marat.

Marat, in the years before the Revolution, had a passion to acquire a scientific reputation. In 1780 he wrote a pamphlet called *Physical Researches on Fire*. In this he claimed that the flame of a candle in an enclosed container went out because hot air pressed on it, that fire was an igneous fluid.

When people ignored Marat's pamphlet he caused a news report to appear in the *Journal de Paris* saying that the Academy of Sciences (of which Lavoisier was president) had given its seal of approval to his theory.

Lavoisier was quick to repudiate this false report. He wrote to the *Journal* that the Academy had *not* approved Marat's theory — which

---

* This list was presented in Lavoisier's *Traité Elementaire de Chimie* of 1789. In addition to oxygen his new elements were sulphur, phosphorus, carbon, antimony, silver, arsenic, cobalt, copper, tin, iron, manganese, mercury, molybdenum, nickel, gold, platinum, lead, tungsten and zinc. Familiar substances like iron, silver and gold had of course been known since remote antiquity, but it was Lavoisier's achievement to recognize them as *elements*, which cannot be broken down into anything else.

** About £500,000 in modern money, in so far as such sums can be compared.

*** This wall of Paris, familiar to readers of *The Scarlet Pimpernel*, was afterwards used by the Revolutionary Government to prevent its victims from escaping. It has been long thrown down, and only the names of its gates remain.

was nothing more than the old phlogiston theory in disguise — and that it had no intention of doing so. And for good measure, he denounced Marat's unscrupulous method of bringing it to public attention.

From that moment on Marat was his deadliest enemy.

Lavoisier, it should be explained, was also an enthusiastic public servant. When the Revolution began in 1789 he was a director — among many other things — of the Powder Commission, responsible for the safe-keeping of gunpowder. Some powder had been stored in the Bastille just before its fall. It was soon alleged that he had done this deliberately, to blow up the 'patriots' who were storming the fortress. His other posts included an association with Necker, King Louis XVI's Finance Minister. He now sought further public office.

Marat meanwhile had started a newspaper called *L'Ami du Peuple* ('The Friend of the People'), which he used to stir up hatred against aristocrats and moderates.* Here is one of his tirades:

And what of Lavoisier, noisy father of all noisy discoveries? He has no ideas of his own, so he appropriates those of others. But since he cannot understand them, he abandons them as easily as he adopts them, changing theories as he changes his shoes. In the space of six months, he has picked up in turn the doctrines of fire, igneous fluid and latent heat. I have seen him first infatuated with phlogiston then ruthlessly abandoning it. Some time ago, following the lead of Cavendish, he discovered the secret of making earth from water.** Then, imagining that this liquid is composed of pure air and inflammable air, he changed it into combustibles.

If you ask me what this man has done to warrant such praise, my reply is that he has got for himself an income of 100,000 livres, has placed Paris in prison with his great wall, has changed the term acid into oxygen, phlogiston into azote, marine into muriatic, nitrous into nitrogen. These are his claims to immortality. Proud of his achievements, he rests on his laurels while his toadies praise him to the skies.

---

* The former 'scientist' Marat had by this time become a feared politician. 'People feared to speak in front of him,' said one of his supporters. 'At the slightest contradiction he shows signs of fury. And if one persists in one's opinion, he will fly into a rage and foam at the mouth.'

** This is a travesty of Lavoisier's actual experiment. He wished to disprove the old belief that water (one of the four legendary elements) could be turned into one of the others, earth. He boiled some water for 100 days in a sealed flask. Sediment at length appeared in the flask. The water did not change its weight, but the flask itself became lighter. The sediment was thus glass from the flask, slowly etched away by the heating. Marat was wrong about Henry Cavendish. He never conducted any such experiment.

I denounce you, Corypheus [chorus leader] of charlatans, Sieur Lavoisier, son of a land-grabber, pupil of the Genevese stockbroker [Necker], tax farmer, Commissioner of Gunpowder, member of the Academy of Sciences. Just to think that this contemptible little man who enjoys an income of 100,000 livres has no other claim to fame than that of having put Paris in prison with a wall costing 30 millions, and that of having transported gunpowder to the Bastille on the night of July 13, a devil's intrigue! And now he seeks to get himself elected administrator of the Department of Paris. Would to heaven he had been strung up to the nearest lamp-post!

Marat was assassinated in 1793 while demanding more executions. But the effect of his smears remained. A year later, at the height of Robespierre's Terror, Lavoisier was arrested with twenty-seven others on charges of having been a tax-farmer — changes which he could not easily deny. J. A. Cochrane tells the story of Lavoisier's 'trial' in his biography:

The accused were surrounded by police with fixed bayonets. The president of the court was Coffinhal, a man of large stature and resounding voice, truly one to inspire terror in the enemies of the Republic. The mob was there in force to gloat over the plight of this bunch of aristocrats. And they enjoyed themselves. Right from the beginning, the answers of the accused to questions put to them by the president were received with derisive laughter. The atmosphere of the court was one of ribaldry.

The public prosecutor opened his case by throwing a series of accusations at the prisoners. One of them was that the Farmers had sent false returns to the State regarding the profits of their contracts in order to get better terms for the next. When one of the defendants replied that the price of each contract was fixed by the Government, and not by the Farm, the president angrily demanded that the answers should be plain yes or no; in other words, they must not argue with the court.

After asking more questions, the prosecutor made his speech in which he accused the prisoners of organised robbery of the State and of being 'the authors of all the evils which for some time had afflicted France.'

The case for the defence was then put, but with the best will in the world the counsel for the prisoners could not in the circumstances have put forward a strong plea, since they had had no time for preparation. But had they the best will in the world? They dared not appear to support the accused too strongly; if they had done so, they would probably have soon found themselves in the dock beside their clients. They brought forward what was at best mitigating circumstances [such as Lavoisier's contributions to science], but these were held to be irrelevant. It was at this point that Coffinhal made his infamous remark: 'The Republic has no need of scientists.' The defence, muzzled and inept, could make no impression on judge or jury, who had made up their minds about the verdict before taking their seats. But the outward forms of justice were gone through, a mere ritual that satisfied the ignorant that all was being done decently, and that satisfied the powers that were in their lust for revenge.

Next came the president's summing up. Coffinhal, himself a lawyer, realised that there was one weak point in the case for the prosecution, that the Revolutionary Tribunal had no power to deal with crimes committed before the Revolution. Sending 28 men to the guillotine when it was beyond his powers to do so might provide an enemy with a lever with which to encompass his own downfall.* He went to considerable pains to lay out the charges in order:

'Has there existed a conspiracy against the French people to favour by every means possible the enemies of France, by exercising all kinds of exactions and extortions from the people, by watering down our tobacco,** by taking six and ten per cent interest for the investment of the funds of the General Farm, while they were legally entitled to only four per cent, by retaining sums which should have been sent to the National Treasury, and by robbing the people and the Treasury by every means possible to deprive the nation of immense sums necessary for the war against the despots arisen against the Republic, and to hand them over to the latter?'

There must be few cases in legal history where the judge in his summing up introduced a new charge against the accused. The fact that the alleged offences had been committed five, 10, 15 years previously, and that the war had only began that year, in 1794, did not affect the case. Coffinhal had now put the issue on a sound legal basis, and he had ensured the conviction of the prisoners. The jury's verdict was unanimous: Guilty.

There followed scenes of enthusiasm among the spectators. Revenge is sweet. Exultant crowds followed the condemned to the Conciergerie prison, where they were handed over to the executioner, Sanson. Bundled into tumbrils, 'biers of the living,' they were slowly jolted through the streets, to the accompaniment of a singing, dancing, shouting throng. Once during the journey the tumbrils were stopped at a certain corner to afford some privileged dwellers to hurl insults at their victims. On again, and so

---

* This in fact happened. The reader will be glad to hear that Coffinhal was executed a few days after the fall of Robespierre.

** Another of Marat's accusations. Lavoisier had been a member of the Farm's tobacco committee for Paris. Marat accused the committee of 'watering down' the tobacco to increase their profits. But in fact the water was added to preserve it while in storage. It was dried out before being sold, and the water could not have affected the taste.

at last, to the accompaniment of the *Carmagnole* [the Revolutionary song and dance], they reached the scaffold.

Here were no formalities: he whose name was first on the list was ordered to mount the scaffold. The knife fell, the severed corpse was removed, and the next on the list took his place. The fourth was Lavoisier.

His death, at the age of fifty-one, was among the most deplorable casualties of the Revolution. The astronomer Joseph Lagrange mourned: 'A moment was all that was necessary to strike off his head. Probably 100 years will not be sufficient to produce another like it.'

Genius Against Stupidity

The attainments of modern science would be impossible without complex algebraic equations. Two statements — and only two — can be made about an equation: either it can be solved or else we can prove it insoluble.

The achievement of the French youth Evariste Galois, whose short, tragic life is described below by the mathematician Eric Temple Bell, was to analyse equations relating to 'groups'. Groups can be quantities of anything, blonde maidens, gray-haired clergymen or professors with horn-rimmed glasses. The only requirement is that they all have to obey the same rules: in this case they all behave as human beings. Galois made crucial discoveries in group theory. Without his work — only recognized for its value long after his death — it would be impossible today, for example, to analyze the configurations of molecules and crystals that are vital in the chemistry of human genes and in the circuitry of computers.

But science, as Bell perhaps fails to make clear, might have advanced even more rapidly than it has if Galois had been less ardently interested in politics, and a less rebellious and difficult student.

*Evariste Galois was done to death by stupidity. He beat out his brief life fighting one unconquerable fool after another.

His mathematical genius came on him like an explosion. In 1823, at the age of twelve, he entered the lycée of Louis-Le-Grand in Paris. The place was dismal. Barred and grilled, it was dominated by a provisor who was more of a political gaoler than a teacher. The France of 1823 still remembered the Revolution. It was a time of plots and counterplots. All this was echoed in the school. Suspecting the provisor of scheming to bring back the Jesuits, the students struck, refusing to chant in chapel. Without even notifying their parents, the provisor expelled those whom he thought most guilty. Galois was not among them, but it would have been better for him if he had been. This tyranny warped one side of his character for life. He was shocked into rage.

But his mathematical interest was already stirring. He began studying mathematics in the regular school course. The splendid geometry of Legendre came his way. Galois read Legendre's book as easily as other boys read a pirate yarn. A single reading sufficed to reveal to him the whole structure of elementary geometry.

Galois's peculiar gift of being able to carry on the most difficult investigations entirely in his head helped him with neither teachers nor examiners. Their insistence upon details which to him were obvious or trivial exasperated him beyond endurance, and he frequently lost his temper. Nevertheless, he carried off the prize in the general examination. His character now underwent a profound change. Knowing his kinship to the great masters of algebraic analysis, he felt an immense pride and longed to rush on to the front rank to match his strength with theirs. He seems to have inspired a mixture of fear and anger in his teachers. They were good men, but stupid, and to Galois stupidity was the unpardonable sin.

At sixteen he was already well started in fundamental discovery. Without preparation, he took the competitive examinations for entrance to the great Ecole Polytechnique. Here, his mathematical talent would be recognised and his craving for popular liberty would be gratified; for were not the audacious young Polytechnicians always a thorn to reactionary schemers who would undo the glorious work of the Revolution?

Galois failed in the examinations. His comrades were stunned. They believed he had mathematical genius of the highest order, and they suspected his examiners of incompetence. Nearly a quarter of a century later Terquem, editor of the *Nouvelles Annales de Mathematiques*, commented on the failure of Galois and on the inscrutable decrees of the Polytechnic examiners. "A candidate of superior intelligence is lost with an examiner of inferior intelligence," says Terquem.

The failure embittered Galois for life.

By 1826 he was fifteen. It was his great year. For the first time he met a man who understood his genius, Louis-Paul-Emile Richard, who kept himself abreast of the progress of living mathematicians to pass it on to his pupils.

Richard recognised instantly the talent of Galois. The original solutions to difficult problems which Galois handed in were proudly explained to the class. Richard declared that this extraordinary pupil should be admitted to the Polytechnique without examination. He gave Galois the first prize and wrote in his term report: "This pupil works only at the most advanced parts of mathematics." It was true. Galois at seventeen was making discoveries of epochal significance in the theory of equations, discoveries whose consequences it took more than a century to exhaust.

The leading French mathematician of the time was Augustin Cauchy. As a rule, Cauchy was a prompt and just referee of the work of others. But occasionally he lapsed. To Cauchy's carelessness, mathematics is indebted for a major disaster. Galois had saved his fundamental discoveries for a memoir to be submitted to the Academy. Cauchy promised to present it, but he forgot, and he lost the author's abstract. That was the last Galois ever heard of Cauchy's generous promise.

Two further disasters in his eighteenth year put the last touches to his character. He presented himself a second time for the entrance examinations at the Polytechnique. Men unworthy to sharpen his pencils sat in judgment on him.

That examination has become a legend. Galois' habit of working almost entirely in his head put him at a disadvantage before a blackboard. Chalk and erasers embarrassed him — till he found a proper use for one of them. During the oral part of the examination one of the inquisitors ventured to argue a difficulty with Galois. The man was both wrong and obstinate. Galois lost all patience. He hurled the eraser at this tormentor's face. The doors of the Polytechnique were closed for ever against him.

After this second failure, Galois returned to school to prepare for a teaching career. The school now had a new director, a cautious royalist supporter. This man's temporising in the political upheaval which was presently to shake France to its foundations had a tragic influence on Galois' last years.

In 1830, at nineteen, Galois was admitted to university standing. Again, his knowledge of his own ability was reflected in contempt for his plodding teachers. During this year he composed three papers which broke new ground. These contain some of his great work on the theory of algebraic equations. Galois hopefully submitted them in a memoir to the Academy of Sciences in competition for the Grand Prize in Mathematics. This prize was still the blue ribbon in mathematical research. Experts agreed that his memoir was more than worthy of the prize. He said with perfect justice: "I have carried out researches which will halt many savants in theirs."

The manuscript reached the Secretary safely. The Secretary took it home with him, but died before he had time to look at it. No trace of it was found among his papers. After Cauchy's lapse, Galois can hardly be blamed for ascribing this new misfortunes to less than blind chance. The thing looked too providential to be a mere accident. His hatred grew, and he flung himself into politics on the side of republicanism.

The first shots of the revolution of 1830 filled Galois with joy.* He tried to lead his fellow-students into the fray, but they hung back, and the temporising school director forbade them to leave the school. Galois tried to escape during the night, but the wall was too high for him to climb. Thereafter, all through "the glorious three days," while the heroic young Polytechnicians were out in the streets making history, the director kept his charges under lock and key. The revolt accomplished, Galois now shocked his family and friends with the fierce championship of the rights of the masses.

The last months of 1830 were as turbulent as is usual after a thorough political stir. Galois contrasted the time-serving vacillations of the director and the feebleness of his fellow-students with the exact opposite at the Polytechnique. He wrote a blistering letter to this effect to the *Gazette des Ecoles*. As a result, he was expelled.

Galois announced the formation of a private weekly class in higher algebra. Here he was at nineteen, a creative mathematician of the first rank, peddling lessons. The course was to have included "a new class of imaginaries (what is now known as 'Galois imaginaries,' of great importance in algebra and the theory of numbers); the theory of solution of equations by roots, and the theory of numbers and elliptic functions treated by pure algebra' — all his own work.

Finding no students, Galois temporarily abandoned mathematics and joined the artillery of the National Guard, two of whose battalions were composed of liberals called "Friends of the People."

He had not yet given up mathematics entirely. In one last effort to gain recognition, he sent a memoir on the general solution of equations — now called the "Galois Theory" — to the Academy of Sciences. Simeon Poisson, of fame in probability, electricity and magnetism, was the referee. He submitted a perfunctory report. He found the memoir 'incomprehensible,' but gave no reason for this strange conclusion. This was the last straw. Galois now devoted all his energies to revolutionary politics. 'If a carcase is needed to arouse the people,' he wrote, 'I will donate mine.'

The ninth of May, 1831, marked the beginning of the end. About two hundred young republicans held a banquet to protest against the royal

---

* This revolution overthrew the last Bourbon king, Charles X, and installed his cousin, the more democratic but less competent Louis Philippe.

order disbanding the artillery which Galois had joined. Toasts were drunk to the Revolutions of 1789 and 1793, to the memory of Robespierre, and to the Revolution of 1830. The whole atmosphere of the gathering was revolutionary and defiant. Galois rose to propose a toast, his glass in one hand, his pocket knife in the other: "To Louis Philippe" — the King. His companions saw the open knife. Interpreting this as a threat against the King's life, they howled their approval. A friend of Galois, seeing notables passing by the open window, implored him to sit down, but the uproar continued. Galois was the hero of the moment, and the artillerists adjourned to the street to celebrate their exuberance by dancing all night. The following day Galois was arrested.

A clever lawyer devised a defence to the effect that Galois had really said: 'To Louis Philippe, *if he turns traitor.*' The open knife was easily explained; Galois had been using it to cut his chicken. The saving clause in his toast, according to his friends who swore they had heard it, was drowned by the applause. But Galois would not claim the saving clause.

During his trial Galois launched into a tirade against political injustice. But both court and jury were moved by his youth. The jury returned a verdict of not guilty. Galois left the courtroom without a word.

He did not keep his freedom long. In less than a month, he was arrested again, this time as a precautionary measure. As a "dangerous radical," he was detained without charge. The government newspapers of all France praised this brilliant coup of the police. They now had "the dangerous *republican*, Evariste Galois," where he could not possibly start a revolution. But they were hard put to find an accusation. They at last succeeded in trumping up a charge. When arrested, Galois had been wearing his artillery uniform. But the artillery had been disbanded. Therefore he was illegally wearing a uniform. This time they convicted him. He got six months.

Discipline in the prison of Sainte-Pelagie was light. The majority of prisoners spent their waking hours promenading in the courtyard or drinking in the canteen. Soon Galois, with his sombre visage, abstemious habits, and perpetual air of concentration, became the butt of the jovial swillers. He was concentrating on his mathematics, but he could not help hearing the taunts hurled at him.

"What! You drink only water? Quit the Republican Party and go back to your mathematics — Without wine and women you'll never be a man!" Provoked, Galois seized a bottle of brandy, not caring what it was, and drank it down. A fellow-prisoner took care of him till he recovered. His humiliation when he realized what he had done devastated him.

At last he was put on parole. He then experienced his only love affair. In this, as in everything else, he was unfortunate. Some "worthless girl" [almost certainly a police agent-provocateur] initiated him. Galois took it violently and was disgusted with love, with himself, and with the girl. To a friend he wrote: "I am disillusioned with everything, even love and fame."

What happened on May 29, 1832, is not definitely known. Galois had run foul of political enemies immediately after his release. These "patriots" were always spoiling for a fight, and it fell to him to accommodate them in a duel. In a "Letter to All Republicans,' he wrote: 'I die the victim of an infamous coquette."

These were the last words he wrote. All night, before writing this letter, he had spent hours feverishly dashing off his scientific last will and testament, writing against time to glean a few of the great things in his teeming mind before the death which he foresaw. Time after time he broke off to scribble in the margin "I have not time; I have not time," and passed on to the next frantically scrawled outline. What he wrote in those last hours before the dawn would keep generations of mathematicians busy. He had found the true solution of a riddle which had tormented them for centuries: under what conditions can an equation be solved? But this was only one thing of many. In this great work, Galois used the theory of groups with brilliant success.

In addition to this distracted letter, Galois entrusted his scientific executor with some of the manuscripts which had been intended for the Academy of Sciences. Fourteen years later, in 1846, Joseph Liouville edited some of the manuscripts for the *Journal de Mathematiques Pures et Appliquées*. Liouville, himself a distinguished and original mathematician, and editor of the *Journal*, writes as follows in his introduction:

"The principal work of Evariste Galois has as its object the conditions of solvability of equations by roots. The author lays the foundations of a general theory which he applies in detail to equations whose degree is a prime number."

Liouville then states that the referees at the Academy had rejected Galois's memoirs on account of their obscurity.

"An exaggerated desire for conciseness was the cause of this defect. One should strive above all else to avoid it when treating the abstract and mysterious matters of pure algebra. Clarity is all the more necessary when one tries to lead the reader farther from the beaten path and into wilder territory. As Descartes said: 'When discussing transcendental questions, be transcendentally clear.' Too often Galois neglected this precept.

"But my zeal was well rewarded when I saw the complete correctness of the method by which Galois proves this beautiful theorem: *In order that an irreducible equation of prime degree be solvable by roots, it is* *necessary and sufficient that all its roots be rational functions of any two of them.*"*

Galois addressed his will to his friend Auguste Chevalier, to whom the world owes its preservation. "My dear friend," he began, "I have made some new discoveries in analysis." He then proceeds to outline such as he has time for. They were epoch-making.

Early on the thirtieth of May, 1832, Galois confronted his adversary on the "field of honour." The duel was with pistols at twenty-five paces. Galois fell, shot through the intestines. No surgeon was present. He was left lying where he had fallen. A passing peasant took him to hospital. Galois knew he was about to die. Inevitable peritonitis set in. His young brother arrived in tears. "Don't cry," Galois said, "I need all my courage to die at twenty."

Early next morning, he died. He was buried in a common ditch, so that today there remains no trace of the grave of Evariste Galois. His enduring monument is his collected works. They fill sixty pages.

Medical Heroes

People who experimented on themselves to discover the nature of diseases, often with fatal consequences, make a little-known but inspiring part of medical history. Lawrence Altman, himself a doctor, tells two horrifying stories of scientists who had the courage to do so.

In Lima, Peru, there stands what may be the only statue in the world of a medical student. It memorializes a young man named Daniel Carrion and the experiment he performed on himself in 1885, an experiment that solved a great mystery about a disease that was killing people in South America. Carrion's research conclusively linked, for the first time, a disease of the skin called *verruga peruana* and another of the blood called Oroya fever (because it had struck workers of the Oroya railway line in the Peruvian Andes).

For years doctors had been trying to find out the cause of puzzling bumps that would erupt on the skin and in the mouths of people living in the steep valleys of the Andean cordillera in Peru, Ecuador and Colombia.

---

*   To put this in plainer language, and somewhat oversimplifying, it is often possible to solve a complex equation by breaking it down into simpler components. In Liouville's time, when mathematics could only be performed by working out the arithmetic, a really complex equation could take a lifetime to solve. Today, of course, computers have removed the need for such elaborate mental labour.

These small bumps, which to victims looked like red warts and to doctors like tumours of blood vessels, were inevitably accompanied by fever and severe joint pain. The rash of bumps was called verruga from the Spanish word for warts, and the sometimes fatal condition was known to have existed in the region for centuries.

A worldwide search for clues to the puzzling disease began, and a Peruvian medical society set up a prize competition to spur interest in the disease. Daniel Carrion, a twenty-six-year-old Peruvian medical student, decided to enter. As a youth, Carrion had often accompanied his uncle on trips through the Andes Mountains, where he had seen verruga sufferers first hand and had been deeply impressed by their affliction. Now in his sixth year of medical training, he had spent the previous three years studying the geographic distribution for the thesis he needed for his medical degree.

He was vaguely aware that some verruga patients developed anaemia, a deficiency of oxygen-carrying haemoglobin in the red blood cells, before they developed the bumps on the skin. He also knew that some doctors believed Oroya fever, a disease of the blood, was actually *verruga peruana*, while others did not believe they were even related. Carrion's chief interest was not in solving this controversy. He wanted to study the evolution of the eruption of the skin bumps to see how the onset of *verruga peruana* differed from that of other diseases such as malaria. By clarifying the earliest phases of the disease, Carrion thought he could help doctors treat patients more effectively.

The more he studied the disease, the more he became convinced that he needed to inject material from a verruga into a healthy person. He might then learn whether the disease could deliberately be given to a human and, if so, document the length of an incubation period and the progression of symptoms. He decided to do the experiment on himself. His friends and professors tried to dissuade him, but Carrion insisted on going ahead.

On August 27, 1885, in a hospital in Lima, Carrion examined a young boy whose skin was affected by the disease. Carrion's professor and three of his assistants were with him. All expressed disapproval of what he was about to do and refused to help. Determined, nonetheless, Carrion took a lancet and drew blood from a verruga over the boy's right eyebrow. Then, unsuccessfully, he tried to inoculate the material into his own arm. At this point, one of Carrion's colleagues overcame his misgivings and helped the medical student finish the inoculation.

On September 21 he recorded his first entry in his diary. He told of feeling a vague discomfort and pains in his left ankle. Two days later, that discomfort had worsened to a high fever and teeth-chattering chills, vomiting, abdominal cramps, and pain in all his bones and joints. He was unable to eat or to quench a strong thirst. By September 26, he could not even maintain his diary, and his classmates assumed the task. Carrion's doctors had no therapy to offer other than herb poultices, comfort, and prayers. Carrion was not dismayed; he believed he would recover. But tests showed that his body had suddenly become alarmingly anaemic, with many millions of fewer red blood cells than normal. We now know they were being destroyed by bacteria whose existence was unknown at the time. The anaemia was so severe that it produced a heart murmur — an abnormal sound usually produced by a disorder of a valve in the heart. Carrion could recognise it himself by listening to the sounds transmitted from his heart to the arteries in his neck.

Carrion's condition weakened each day, but his mind remained alert. As he lay sweating and feverish, he began to understand the implications of what he had done. He remembered the varying theories about the links between Oroya fever and *verruga peruana* and recalled the recent death of a friend. On October 1, He told his friends: "Up to today, I thought I was only in the invasive stage of the verruga as a consequence of my inoculation, that is, in that period of anaemia that precedes the eruption. But now I am deeply convinced that I am suffering from the fever that killed our friend. Therefore, this is the evident proof that Oroya fever and the verruga have the same origin.'

Carrion was right. He had shown that *verruga peruana* and Oroya fever were in fact one disease. Whatever caused the mild skin condition of *verruga peruana* could also cause the high temperatures, bone pain, and fatal anaemia or Oroya fever. It was yet to be discovered that both are manifestations caused by a bacterium which is spread by sandflies.

His condition rapidly worsened, and he was finally admitted to the same hospital where he had first inoculated himself. He went into a coma and died on October 5 — thirty-nine days after performing his self-experiment.

Although Carrion was eulogized profusely after his death, at least one prominent doctor publicly criticised the experiment as a "horrible act" by a "naive young man" that "disgraced the profession." And there were some who said he had committed suicide. To complicate the situation, when the police learned the identity of the physician who had helped inject the verruga material into Carrion's arm, they charged him with murder. Carrion's professor came to the defence of his student and the assistant. The professor cited the many physicians in other countries who had risked their lives in self-experiments. As a result of his arguments, the murder charge was dropped. Today, Carrion is an unqualified hero in Peru, where medical students sing a ballad to his memory.

Another physician who used himself as a volunteer for his own experiment was John Hunter, a surgeon to King George III and one of the most celebrated anatomists and medical teachers of his day.

He pioneered in transplant surgery by placing a human tooth in a cock's comb, made some of the finest, most detailed anatomical descriptions of the body's lymph system, and helped us understand how bone grow.

In 1767, Hunter contracted gonorrhea and syphilis, but not in the usual way. Then as now, these two diseases were among the most common ailments that doctors treated, and in Hunter's era there were almost as many theories about their cause as there were physicians. To understand gonorrhea, Hunter took pus from an infected patient and injected it into two places on his own penis in expectation of producing the same infection in himself. Two days after inoculation, the injected areas began to itch, redden, and become moist. A week later, pus formed. There was irritation on urinating. The thirty-nine-year-old surgeon had been success-ful; he had contracted gonorrhea.

Unexpectedly, however, a few weeks later, the characteristic sores of syphilis also developed. Hunter treated himself with calomel ointment and other standard but useless remedies. The syphilitic sores "healed" or, as we now know, naturally disappeared for a time. Four months later the sore on the tip of the penis recurred. Once again it healed on its own. It was to reappear and heal several more times. Meanwhile, a lymph gland in Hunter's right groin swelled. He treated the swelling by rubbing mercury on the leg and thigh as an experimental "cure." The swelling subsided considerably. When it twice returned, he applied increasingly larger doses of mercury for symptomatic relief.

Unfortunately for Hunter, the patient from whom the venereal pus was taken probably had a dual infection — both gonorrhea and syphilis. We now know that syphilis, over the span of decades, passes through three stages and ultimately damages many tissues, such as the heart and the aorta, the main artery leading from the heart. Also, long-term syphilis can be one of the several causes of angina, a painful condition that results when the heart is deprived of an adequate supply of oxygenated blood. Hunter suffered from attacks of angina for the last 15 years of his life. The least exertion, physical or emotional, was apt to induce spasms ending in unconsciousness. Often, the mere act of undressing at night precipitated an attack, as did heated discussions, drinking wine, performing a difficult operation, concern over an experiment in progress, or climbing stairs. As he said: "My life is in the hands of any rascal who chooses to annoy and tease me."

Indeed, his death at the age of 65 followed an argument during a meeting at St. George's Hospital in London, where he practised. An autopsy showed that he had a ballooning of the aorta known as an aneurysm. Modern doctors would attribute his death to the third stage of syphilis.*

---

*   Sadly, Hunter's heroic action does not seem to have accomplished anything. Indeed, he is said to have *set back* progress into the research of venereal disease by a century because he drew false conclusions from his experiment. He died unaware that he had been suffering, not from one disease, but two.

7
Bogus Science
The False Authority

Until the sixteenth century, most people's idea of how to solve a scientific problem was to look up what Aristotle had written about it. His opinions were invested with an almost divine authority, and the orthodox were fond of using them to suppress genuine breakthroughs in science. And yet his opinions were largely rubbish, as the Nobel Prize-winning biologist Sir Peter Medawar and his wife Jean explain:

"Aristotle was 'a man of science' in the modern sense. He was a careful collector and observer of an enormous range of facts . . . Much of his work is still regarded with respect by scientists who care to study it." These two sentences by the humanist Goldsworthy Dickinson betray an almost majestic incomprehension of the character of science and of Aristotle's influence on it. A scientist is no more a collector and classifier of facts than a historian is a man who compiles and classifies a chronology of the dates of great battles and major discoveries.

Although Aristotle's philosophical opinions command respect, the pioneers of or spokesmen for the new science of the 17th century (men such as Robert Boyle of *The Sceptical Chymist*, Dr. Joseph Glanvill, Francis Bacon and the poet Abraham Cowley) repudiated his doctrinal authority with weary exasperation. No one did more than Bacon to bring the doctrinal tyranny of Aristotle to an end. Cowley wrote of his 'pernicious opinion that all things to be searched in nature had already been found and discovered by the ancients,' and Henry Power, declaring that the new science was 'coming in on a spring tide,' wrote: 'I think I see how all the old rubbish must be thrown away.'

It was Aristotle who prompted Bacon to say that *works*, not words, were to carry the message of the new science; and Aristotle's were doubtless the words that persuaded the oldest and most famous scientific society in the world, the Royal Society, to adopt as its motto, *Nullius in verba* — Don't take anybody's word for it.

Aristotle's biological works, in his *Historia Animalium*, are a strange and tiresome farrago of hearsay, imperfect observation, wishful thinking and downright gullibility. What evidence can have convinced him that the semen of youths between puberty and the age of 21 is 'devoid of fecundity,' and that young men and women produce undersized and imperfect progeny? Prolapse of the uterus and menstruation as often as three times within a month are said to be symptoms of 'excessive desire.'

It is hard not to sympathize with those 17th century scientists who resented and repudiated such a doctrinal authority. For poetic truth in Aristotle's conception was a revelation of the ideal, of what *ought to be*. And for him, poetry was a more philosophical and a higher thing than history and science. Poetry alone could reveal what ought to be in the light of an understanding of Nature's true intentions.

There can be no little doubt that Aristotle extended this idea to science — and in this light we can understand some of his bizarre opinions. He was a firm believer, for example, in the Hebdomadal rule, that everthing goes in sevens. Man has seven ages, each seven years long, so anyone with a true understanding of nature would realize that human semen must be infertile between the ages of 14 and 21, for would not fertility in this period contravene the Hebdomadal rule? And so what ought to be became what was. Sometimes of course Aristotle is right; his writings were so voluminous he could hardly fail to be correct sometimes. He thus roundly declares that the semen of the Aethiops is not black and chides Herodotus for thinking that it is.

We do not believe, in short, that anyone who decides not to read the works of Aristotle the biologist will risk spiritual impoverishment.

'The Hollow Earth'

This amazing story is attested to by many authors. The following account is from Bergier and Pauwel's *The Morning of the Magicians*:

We are in April 1942. Germany is putting her whole strength into the war. Nothing, it would seem, could distract the technicians, scientists and military chiefs from the performance of their immediate tasks.

Nevertheless, an expedition organized with the approval of Goering, Himmler and Hitler set out from the Reich surrounded by the greatest secrecy. The members of this expedition were some of the greatest experts on radar. Under the direction of Dr Heinz Fischer, well known for his work on infrared rays, they disembarked on the island of Rügen in the Baltic. The expedition was equipped with the most up-to-date radar apparatus, despite the fact that these instruments were still rare at that time, and needed in the principal nerve-centres of the German defence system.

However, the observations to be carried out on Rügen were considered by the Admiralty General Staff to be of capital importance for the offensive which Hitler was preparing to launch on every front.

Immediately on arrival at their destination Dr Fischer aimed his radar at the sky at an angle of 45 degrees. There appeared to be nothing to detect in that particular direction. The other members of the expedition thought that a test was being carried out. They did not know what was expected of them; the object of these experiments would be revealed to them later.

To their amazement, the radar remained fixed in the same position for several days. It was then that they learned the reason: Hitler had formed the idea that the Earth is not convex but concave. We are not living on the outside of the globe, but inside it. Our position is comparable to that of flies walking about inside a round bowl. The object of the expedition was to demonstrate this truth scientifically. By the reflection of radar rays travelling in a straight line it would be possible to obtain an image of points situated at a great distance inside the sphere. The expedition also had a second object, namely, to obtain an image of the British Fleet at Scapa Flow.

The Lunatic Fringe

That great science writer Arthur C. Clarke once received a letter from a man claiming to have been visited by beings from space, who came from the planet Ying, 50,000 light-years away.

Clarke, an expert at dealing with cranks, ruminated over a reply for some minutes. Then he wrote:

Dear Sir,

You have been completely deceived; the visitors from space who landed in your back garden and informed you that they had come from the planet Ying, 50,000 light-years away, are imposters. I have definite proof that they are actually from the planet Yang, which is only 40,000 light-years away.

Yours faithfully,

Yet Clarke does not consider cranks to be harmless or amusing. He sees them as a threat to freedom. He makes this clear in the following eloquent chapter from his 1965 book *Voices from the Sky*:

The lunatic fringe has always been with us. In every age there have been people who have been willing to believe anything so long as it was sufficiently improbable. Religion, economics, science, politics have all had — and still have had — their fanatical minorities who devote their fortunes, their energies, and often their lives to the cause they have made their own.

Often the cause is a sensible one but its advocates are not. They show that humourless monomania, that inability to see any other point of view, that distinguishes the crank from the enthusiast. One does not have to look very far for examples. The best publicized (perhaps over-publicized) specimens in the United States at the moment are probably the John Birch Society and the Black Muslims.

The driving force behind all such extremist groups and crackpot organizations is a mixture of fear and ignorance. It may be, as in the above cases, a well-justified fear of the Communists or the Ku Klux Klan, but often it stems from deeper and less rational causes. We can see this very clearly if we look at two of the most famous examples of mass moronity in the past decade — Bridey Murphy and the flying saucers. (If Hollywood fancies this title, it can have it.)

In the Bridey Murphy case, a Colorado housewife 'remembered' her life as a girl in Ireland more than a century before, and gave an elaborate account of it while under hypnosis. This was built up, by skilful publicity, as evidence of reincarnation, despite the fact that a little careful research would have revealed the truth. When a few sceptical journalists did this research, and uncovered the childhood sources from which the subject obtained her memories, the whole sensation collapsed almost overnight.

Let me make one point quite clear. The Bridey Murphy affair involved a perfectly genuine and still unexplained phenomenon, almost as remarkable, in its way, as true reincarnation. But this phenomenon — hyperamnesia, or the incredible detailed and *creative* recall of long-forgotten memories under hypnosis — has been familiar to all psychologists since at least the time of Freud. To have placed it before the American public as proof of survival after death was an act of irresponsible incompetence. Some would use stronger terms.

Yet the public lapped it up. The book became, not merely a best seller, but *number one best seller*, and several hundred thousand copies were soon in circulation. The publishing trade can take little pride in such exploitation of fear and ignorance — in this case, fear of death and ignorance of psychology.

The flying saucer craze lasted much longer and indeed is still with us, so there is no need to go into details. But once again, as with Bridey Murphy, we must distinguish between a real phenomenon and the conclusions drawn from it by anxious and hysterical people.

'Unidentified flying objects,' to give them their non-committal name are quite common. If you have never seen a UFO, you should be ashamed of yourself, for it means that you are not very observant. (I've encountered seven, including two that would have convinced the most sceptical.) They have dozens of causes, many of them ludicrously simple, for it is amazing what nature can contrive when she is in the mood — look at the rainbow or the snowflake. A small proportion of UFOs has never been satisfactorily explained, and the theory that they are from outer space is a perfectly reasonable one. I would be the last to condemn it, since I have spent most of my life expounding the possibility.

What I *am* condemning is the credulous naivety of those who have accepted this theory and made almost a religion of it. On the strength of a few faked photographs and the ravings of obviously psychopathic personalities, thousands were convinced that men from space had actually landed on this Earth. Many still believe this, despite the fact that ever since the opening of the International Geophysical Year in 1959, the skies of our planet have been raked by armies of trained observers and every conceivable type of detecting equipment.

The chance of a genuine spaceship evading discovery in this age of multi-billion dollar radar networks, is about the same as that of a dinosaur concealing itself in Manhattan.* When our stellar neighbours really do

---

* I once saw a movie in which this happened. It contained the immortal line: 'It's hiding somewhere in the Wall Street area!' — A.C.C.

start to arrive, we'll all know about it within five minutes. The idea that any government could — or would — keep such a world-shattering event secret *for year after year* is utterly ludicrous.

The fears of the UFO-ologists are more complex than those of the Bridey Murphy believers. (I would guess, by the way, that the two groups overlapped to a very large extent, for credulity knows no boundaries.) Alarm at the drift to atomic destruction was combined with the hope that benevolent saviours from the sky would arrive and tidy up the mess we have made of this planet. And in this case, the ignorance which made so many honest people misinterpret the evidence of their own eyes was completely excusable. Not even the scientists had realized what an extraordinary collection of optical, astronomical, meteorological and electrical apparitions inhabit our skies. The UFOs have done some good by focusing attention on these.

It is depressing to make a list of the other pseudo-scientific ideas that have achieved fame or notoriety during the last generation. Do you remember Immanuel Velikovsky's *Worlds in Collision*? This monument of impressive scholarship explained Biblical history on the theory that the Earth and Venus had cannoned together in the past like billiard balls. L. Ron Hubbard's *Dianetics* belonged to the same category. It purported to cure all mental ills by taking the patient back to the time when his troubles started. Another revelation must be imminent, now that the Beats are buried and the zest for Zen is flagging. I have no idea what it will be, and I am in no hurry to find out.

You may feel that this is making too much of something that affects only a small part (one hopes) of the total population. It is true that in the past crankiness and eccentricity did little harm, and even added a little spice to society. A generation ago, flat-Earthers, end-of-the-world cultists, and disciples of weird religions caused no embarrassment outside their immediate circle. But we are moving now into a complex and perilous age, where credulity and superstition are luxuries that can no longer be afforded. For consider this example:

In 1843, fifty thousand followers of the prophet William Miller gathered on New England hilltops to await the expected hour of judgment. The advent of a great comet, its tail streaming like a fiery banner across the sky, seemed to them a sign that the end of the world was at hand.

People are still watching the sky for signs of doom; but now they look into radar screens. And here is the important difference: the beliefs of fifty thousand Millerites could have no influence, one way or the other, upon the end of the world, but today, when we carry the power of Vesuvius in a single warhead, the fears or delusions of only fifty men could bring it about.

This is an extreme case. But all forms of irrationality are dangerous, because in the right circumstances they can spread like a plague, infecting

161

not only a community but an entire nation. Those concerned may be very ashamed of themselves afterwards, but by then the damage may be done.

You cannot build an informed democracy out of people who will believe in little green men from Venus. Credulity — willingness to accept unsupported statements without demanding proof — is the greatest ally of the dictator and the demagogue. It is not so very long ago that there were voices crying: 'The Jews are plotting against the Reich!' and, 'I have here in my hand a list of 205 Communists in the State Department.' Those voices are silent now, but there will be others.

One of the factors, ironically enough, which has contributed to popular willingness to accept the incredible is the success of modern science. Because so many technical marvels have been achieved, the public believes that the scientist is a magician who can make *anything* happen. It does not know where to draw the line between the possible, the plausible, the improbable and the frankly absurd. Admittedly this is often extremely difficult, and even the experts sometimes fall flat on their faces. But usually, all that is required is a little common sense.

Unfortunately, common sense has always been rather rare. As a reminder of this, let me quote two final examples of mass stupidity, which may also help to dispel the idea that it is an American monopoly.

During the darkest days of the First World War, the rumour swept the length and breadth of Britain that troops were arriving from Russia in huge numbers (this was before the Revolution) to reinforce the crumbling Western Front. Thousands of honest Britons 'saw' them at ports and railway stations, and millions believed the rumour, because they wanted to. And how did the observers know that these soldiers were actually Russians? Not because they said so — but *because they had snow on their boots.*

That little detail was the clincher, as far as most people were concerned. They never stopped to ask if even Russian snow would survive the long sea voyage from Murmansk to Scotland.

My last example may surprise you. You may not know that flying saucers have invaded the Soviet Union. Yet they have, for a remarkable reason. According to *Pravda*, which is rather indignant about the whole affair, Russian saucer fans believe that little people from Venus constantly descend on Uzbekistan, Tajikistan and then 'promptly scurry in all directions in search of inexpensive Oriental sweets.'

I love that 'inexpensive'; presumably the ruble is hard to get on the Venus black market.

No one should derive much satisfaction from this proof that nuttiness is also rampant on the other side of the Iron Curtain. Unreason is always a menace, wherever it occurs. It may be even more of a danger in the Soviet Union than in a country with democratic safeguards. (Look at what Hitler's intuition did to the world.) And there is, unfortunately, no reliable cure for it. You cannot buy sanity at the chemist's, or inject common sense by mass inoculation.

The only answer lies in education, and even that is merely a palliative, not a panacea, for a college degree is no guarantee of wisdom, as anyone who has ever been near a campus will testify. There are many people in the world who have been educated beyond their intelligence, but there are far, far more who have not been educated to within hailing distance of it. They are the ones who provide fodder for the demagogues and cranks, who listen to false prophets and sponsor absurd or evil causes. They cannot always be blamed for society has robbed them of what should be everyone's right — an education to the limit of their ability, whatever their financial status, creed or colour. No wonder that, dimly realizing their deprivation, they seek any substitute that they can find.

Very often that substitute takes the form of anti-intellectualism — a pretence that knowledge, education and culture are worthless or even dangerous. This is, of course, a typical sour grapes reaction. Not long ago, one could identify those suffering from it by their fondness for the phrase 'egg head.' That engaging term is now a little out of fashion, because the events of the last few years have made it obvious to everyone that a society that despises brains is on the one-way road to oblivion.

Human nature being what it is, the lunatic fringe can never be abolished, and most of us, if the truth be told, would hate to see it vanish altogether. But education can minimize its influence, can convert it from a potential danger to a source of mild amusement. A century ago, Matthew Arnold compared this world to a 'darkling plain . . . where ignorant armies clash by night.' The metaphor is still valid. Perhaps the greatest single task that now faces every nation is the dispelling of that ignorance, lest the armies clash again — for the last time.

A Bloody Bore

Research in the paranormal does not lack for apparent respectability. The American Parapsychology Association is a division in good standing of the American Association for the Advancement of Science. Several leading universities, including the University of California at Davis, have professors of parapsychology, and Ph.D.s in the subject are granted. Although such research has yet to produce anything in the way of a repeatable controlled experiment, its practitioners argue that its revolutionary potentialities justify its continuation. My own feeling is that after a century of total failure it has become a bloody bore.

— Dennis Flanagan

The Bogus Dentist

Scientific charlatans like William Fuller will always be with us. The famous magician James Randi explains why they are so dangerous:

Faith-healing abounds with ridiculous claims, but none sillier or more preposterous than those we will now examine. The Reverend Willard Fuller, of Palatka, Florida, says he can insert dental fillings without drillings, or even opening his client's mouth, turn ordinary silver fillings and crowns into gold, straighten crooked teeth, tighten dentures, cure periodontal disease, and grow new teeth in his clients — all just by calling on Jesus to do it. As he puts it:

> Sometimes you can watch a cavity fill right up in front of your eyes. You can actually see silver, gold or porcelain coming up until the whole cavity is full. It's amazing!

Fuller calls himself the 'Psychic Dentist.' Believers swear that all the above-listed miracles have taken place in their mouths at his command, though none of them have produced before-and-after X-rays to prove these claims. Much of Fuller's success depends upon the fact that most people do not know where fillings or other repairs are actually located in their own mouths. The faithful also find wonders in every little twinge and tingle they experience while in the ecstasies of evangelical fervour. Add a little carefully applied hoopla and tambourine-thumping, and you have dental miracles.

The Fuller literature abounds with fantastic claims. It is easy to simply spout accounts of miracles when they cannot be checked out. In a meeting in Rochester, New York, in October, 1986, Fuller claimed that an 11-year-old girl in Phoenix, Arizona, who had six cavities suddenly had no cavities and no fillings after he cured her. A 66-year-old woman 'in upstate New York,' he said, had *no teeth*. But with Fuller's magic, so he claimed, she 'began cutting a set of baby teeth, and eventually re-grew an entire new set of teeth.'

At that same meeting, Fuller's wife Amelia declared: 'We welcome sceptics with an open mind who come to investigate.'

Acting upon that invitation, Mark Plummer, of the Committee for the Scientific Investigation of Claims of the Paranormal, tried to obtain the identities of the Phoenix girl and the woman with the set of new teeth. The Fullers declined to supply that information.

(On December 4, 1986, Mr Randi attempted to file criminal charges against Willard Fuller with the Florida state government. Eight months later, he was still awaiting a decision from the appropriate regulatory department.)

Only on one occasion did the authorities try to check out Fuller's claims. When he and his wife visited Australia early in 1986, the Dental Board of New South Wales charged him with breaching the Dental Act by falsely advertising himself as a dentist. He was convicted and ordered to pay court costs of 435 Australian dollars. But, says Fuller, because of his saintly nature and his religious orientation, the conviction was 'not recorded'.

Back in the United States, Fuller began rewriting history. In the July 1, 1986, issue of his newsletter, the *Lively Stones Fellowship*, he exulted that 'we issued the order for prayer power and for 150 angels to march into the courtroom with us.' He said he had been 'invited to court' by the Dental Board, that 'the case was dropped,' and that the judge read a two-page statement saying that the character of the Fullers and their ministry was 'beyond reproach.'

These quotes were inventions. Fuller was not 'invited' to court; he was 'commanded to appear under penalty of law'. The case was not 'dropped'. It was heard in all its detail and proved. The judge never declared on his character; the words Fuller quotes never appeared anywhere except in his newsletter. And as for the crowd of angels in the courtroom, they were unnoticed by anyone, nor were they reported in the Press.

How can Fuller's victims possibly believe that God puts gold and porcelain crowns and silver fillings into their mouths? Why doesn't He just restore the original tooth? After all, He made the original tooth, didn't He? Fuller is credited with a 'miracle' when people see in their mouth silver fillings that have turned to gold. He uses a small, penlike torch to look into the victim's mouth. Its feeble glow is rather yellowish, and could make silver fillings appear golden. In fact, many recipients of this Midas touch complain that the next day the miracle is found to have reversed itself, the gold having turned back into silver.

How does Fuller claim to perform this magic? His explanation is typically crackpot. Basic, accepted facts are followed by sheer nonsense. As he says:

> Everything in the Universe, including our bodies, is made up of atoms. The atoms can be manipulated, and when you get into the right relationship with God, you have a great source of power at your disposal.

The dangers of Fuller's operation are many. As I vainly informed the Florida Department of Professional Regulation, he wanders about his audience poking a dental mirror into mouth after mouth, swishing it in a glass of something or other between pokings. Such a procedure is certainly insufficient to sterilize the instrument, and it could spread deadly diseases such as the AIDS virus from one infected person to all others he touches.

Why are people like Fuller so popular? Chris Florentz, director of communications of the Dental Society of the State of New York, believes they may attract dental phobics — people who will avoid receiving dental care in almost any circumstances.

But Florentz was pessimistic about the possibility of prosecuting the Fullers. 'The First Amendment (which guarantees free speech) and freedom of religion adds to the complexity of the issue,' he says.

Yes, and only after some serious illnesses have occurred from infections caused by the Reverend Fuller will anyone trouble to defend us from this perversion of the First Amendment.

I have only one simple question for the Reverend Fuller. How is it that he has six missing teeth himself, while the rest are badly stained and contain quite ordinary silver fillings? Physician, heal thyself.

8
Utilities

Kaldi and the Coffee Bean

Millions of coffee-drinkers should be grateful to Kaldi, the Yemeni goatherd, who discovered the powers of the coffee bean in about A.D. 850. The story was painstakingly gathered by Antonius Nairone, a seventeenth-century Maronite monk and scholar, and professor of theology at the Sorbonne. Henry Jacob tells it:

Between men and goats there has always been a bargain. Such was the case here, in the Yemen desert by the shores of the Red Sea. There was scant vegetation among these foothills. Across them one glimpsed rust-coloured mountains. Nothing grew on their heights, and men did not visit them. Only the runaway goats, from time to time, climbed the topmost peaks. After weeks, they would return, lean and out of condition.

The goats supplied the nearby Shehodet Monastery with milk and goat-hair; while the monks of the monastery protected the goats with herdsmen and watchdogs.

The goatherds knew their charges well. They knew the creatures' fondness for climbing, and for gnawing the barks off trees; their perpetual craving for salt. They knew that goats would often take to the high mountains for a week or more, and be slow to return. But now the beasts were displaying a new characteristic which troubled their keepers. They were affected with sleeplessness.

For several nights in succession, they clambered over rocks, cutting capers, chasing one another, bleating fantastically. With reddened eyes they gambolled convulsively when they caught sight of the goatherds, and darted away madly.

They could not explain it and the chief of the monastery, the imam, came to investigate. Two of the animals were brought to him. There was no sign that any bird had been pecking their teats, as the goatherds suggested.

'Your goats must have eaten poison,' declared the imam.

'What poison could they have found?' demanded the chief goatherd, a man named Kaldi.

'Follow them, and keep them under close watch.'

The goats were as active as before, and they did not sleep.

At last Kaldi held out a spray.

'We've found the plant that has bewitched them!'

It was that of a shrub rather than a tree, with dark-green, firm and tiny leaves, somewhat resembling those of a laurel. Short, white blossoms, projected from the leaves, closely set like those of jasmine. Some had ripened and fallen, leaving their fruit, little berries of a strange, violet hue. If you grasped one of these, between finger and thumb, you could feel that it had a thick, hard kernel. The imam turned the spray over and over, much astonished. What he held resembled the character of many plants he knew, but in its assemblage of characters unknown.

'Your goats have been eating this?'

Kaldi replied that a coppice of the unknown shrub had been discovered, obviously devastated by the goats.

'In what direction?'

'Towards the north.'

After climbing for several hours over boulders and brambles, the imam and Kaldi, guided by the other goatherds, reached the coppice. It was in a wady, damp and hot. What remarkable trees! They ranged in height from six to twelve feet, and looked more like overgrown shrubs than trees.

No one had ever seen anything of the kind before. The imam plucked some of the leaves and blossoms and chewed them, but soon spat them forth. They had no taste worth speaking of, nor any scent that could have allured the goats.

Back at the monastery, the monastery's parchment volumes were searched in vain for a description of the unfamiliar plant.

'To my way of thinking,' said the imam, 'this shrub is not a wild one but a cultivated one that has escaped from a garden and run wild.'

'How could there have been a garden in or near such a desolate spot?' asked Kaldi.

'You must have heard that, centuries back, our land was conquered by the giaours. I mean the black Christians from Africa, subjects of the Monarch of Ethiopia. They crossed the Red Sea, coming from the territory known as Kaffa. They brought with them their favourite vegetables and flowers. I think this is a Kaffa tree —'

'If the tree has magic virtues, surely we should have heard of it,' said Kaldi dubiously. 'It is but a tree like many another.'

That night the imam was determined to discover whether the strange plant really had mysterious powers.

He decided to make an infusion.

He crushed some of the leaves and flowers in a glass vessel with a spoon. His spoon encountered the kernel, which contained both berries and hard beans. He turned his attention to the beans. He boiled some water and threw it over them. The result was a dark brew, while an aroma rose from the pot such as had never before assailed his nostrils.

The imam drank a glassful of the steaming decoction. It was bitter to the taste, savouring of charred wood. Then he lay down and composed himself to sleep.

Within a few moments, he began to feel as if under a spell. He was in a state of intoxication.

He was aware of the quickening beating of his heart. He was sweating a little, and had a wondrous sense of lightness in the limbs. His mind was unusually active, cheerful and alert. He felt as lively and vigorous as if he had been refreshed by thirty hours sleep. Springing up, he began to pace the chamber untiringly.

At midnight, the hour for prayer, he went along the corridors to awaken the brethren in their cells. Heavy with sleep, they reluctantly arose, as commanded by the Prophet.

But the imam came to each in turn, and they moistened their lips and tongues with the black, bitter drink. It had an unpleasant taste although the odour was pleasant enough. It gave them the will to wake. One who swallowed enough of it forgot that he had been prematurely aroused from slumber. The sense of fatigue departed from his knee-joints, and he no longer felt the dead weight of his arms hanging from his shoulders.

Night after night, at the hour of midnight prayer, the imam and his monks refreshed themselves with the decoction of the Kaffa-seeds. In their thankfulness, they gave the elixir a name with a twofold meaning. They called it 'Kahveh,' the stimulating, the invigorating; this described the magical qualities of the coffee-bean as well as its supposed derivation from Kaffa. They thought also of Kawus Kai, the great King of Persia, who had conquered the force of terrestial gravity and had ascended into heaven in a winged chariot.

The Reckoning Master

The good Christian should beware of mathematicians, and all those who make empty prophecies. The danger already exists that mathematicians have made a covenant with the Devil to darken the spirit and to confine man in the bonds of Hell.
—St Augustine in the fifth century A.D.

The hostility of the early Christians to science was intolerable to the thinkers of the Renaissance. This was particularly true in the case of practical mathematics, as Frank J. Swetz explains:

By the beginning of the 16th century, there was a lag of perhaps 200 years between the commercial and financial methods of the Hanseatic League merchants and their more advanced southern counterparts. From the 14th century on, merchants from the north travelled to Italy, particularly to Venice, to learn the mercantile arts of the Italians, notably in commercial arithmetic.

On acquiring these skills, they returned home with a new Italian vocabulary that included such terms as: *disagio*, discount; *credito*, credit; *valuta*, value; and *netto*, at a net price. These words found their way into contemporary business jargon. Even Jacob Fugger, the great German merchant prince, left Augsburg to study business technqiues in Venice. The flow of German merchants seeking knowledge was so great that the Venetians established a special institution to accommodate them. The 'German Factory,' a five-storey structure with a great courtyard, accommodated over 80 visiting merchants and their servants and served as a warehouse, hotel and market. The skill they all desired was a proficiency in Italian commercial arithmetic.

Early during their rise to commercial supremacy, the Italians, and particularly the Venetians, realized the importance of arithmetic. In contacts around the Mediterranean and Barbary coasts, Italian merchants became exposed to the Hindu–Arabic numeral system and its methods of computation. The mathematician Leonardo Fibonacci studied this new arithmetic under an Arab master.* He became convinced that the new numerals were vastly superior to the Roman numerals commonly used in Europe. The evangelist of this new knowledge, he published in 1202 a book, *Liber abaci*, of which a section was devoted to the commercial applications of arithmetic. The message was well-received in the merchant houses of Pisa, Genoa and Venice. Hindu–Arabic symbols soon began replacing Roman numerals as the use of the abacus gave way to computations with pen and ink.

The Italians not only adapted the new methods for their commerce, they also became innovators — the practice of double entry bookkeeping originated in Italy at this time. In this technique, each entry is recorded twice, on one page under a debit heading and opposite under a credit title.

* Fibonacci (1180–1250) became belatedly famous in the twentieth century for his discovery of the Fibonacci numbers, a series such as 1,1,2,3,5,8,13, etc. where each number is the sum of the previous two. It has many applications in the real world, from flower petals to the orbit of planets — A.B.

By 1400 this method of accounting was well established in Italy as a basic financial tool. Venetian ledgers were admired and copied by visiting foreign merchants. Venice created a university chair of mathematics devoted to navigation. Piero de Versi, an early holder of it, produced one of the first tracts teaching both mathematics and navigation.* And Venice became the first European city to endow public lectures on algebra.

Italian mathematical education became so excellent that it attracted students from all over Europe, as Tobias Dantzig related:

A story of a German merchant of the 15th century is typical of the situation then existing. Desiring to give his son an advanced commercial education, he appealed to a prominent university professor for advice. The reply was that if the young man wanted only to learn to add and subtract, he could obtain this instruction in Germany. But Italy was the only country where he could learn to multiply and divide.

But even in Italy, if one wished to study numerical computing, particularly for commerce, one did not necessarily do so at a university. By the mid-13th century, arithmetic was still being taught as a science at European universities. Yet the level of teaching was often theoretical and without practical applications. Under the influence of scholasticism its usefulness was held to be questionable compared with grammar, logic and rhetoric. In Henri d'Andeli's *Battle of the Seven Arts*, a commentary on the teaching in medieval universities, the author depicts arithmetic as an aloof maiden detached from the reality surrounding her:

Arithmetic sat in the shade,
Where she says, and she figures,
That ten and two and one make thirteen,
And three more make sixteen;
Four and three and nine to boot
Again make sixteen in their way;
Thirteen and twenty-seven make forty;
and three times twenty by themselves
make sixty.
Five twenties make a hundred, and ten
hundreds a thousand.

---

* *Alcune raxion dei marineri* (1444).

Does counting involve anything further?
No.
One can easily count a thousand thousands
In the foregoing manner.
From the number which increases and
diminishes
And which in counting goes from one to
hundred.
The dame makes from this her tale,
That usurer, prince and count
Today love the countress better
Than the chanting of High Mass.

In universities that taught both practical and theoretical arithmetic, the fees for studying the latter often doubled those for the former, showing the perceived discrepancy between the value of the two subjects. If a student in the early Renaissance wished to learn commercial mathematics, he usually did not go to a university at all, but sought out a 'reckoning master,' a man skilled in the arts of commercial computation.

Europe's mercantile development from the 13th century onward demanded increased proficiency in commercial arithmetic. Merchants found themselves obliged so to instruct their sons and apprentices. As the tempo of trade increased, they had to hire reckoning masters. They were well-respected for their computational skills and much sought after. Cities even passed ordinances on how far apart they could live. The profession of reckoning master became lucrative, and often they banded together to form guilds and associations. Many of them were self-employed, a sort of mathematics consultants-at-large. amd they accepted students for private tuition or conducted group classes. They, and reckoning schools, rapidly proliferated in the commercial cities and along the trade routes of Europe. In 1338, Florence had six such schools, and by 1316 Nuremberg alone had 48 of them.

Students of the reckoning schools were usually sons of merchants or civil servants, children of the middle class, probably aged between twelve and sixteen. Young merchants travelled to them from all over Europe to learn the *Welsche Praktik*, the mathematical arts, of their trade. The seriousness of this task is emphasized by the advice of a father to his son studying in Venice, to 'rise early, go to church regularly and to pay attention to his arithmetic teacher.'

The Light-Mile

So minutes, hours, days, months and years,
Passed over to the end they were created.

Henry VI, Part 3

There is no mention here of seconds. Indeed, they were not used until 1760, when John Harrison invented his marine chronometer for accurate navigation. Today, with the discovery of ever smaller sub-atomic particles, it has been necessary to use ever smaller units of time.

Many of these have been clumsy and confusing. Here Isaac Asimov proposes a brilliant new method of measuring micro-time, based on the speed of light:

If we are permitted to speak of a light-second as equal to the distance covered by light (in a vacuum) in one second and find it equivalent to 186,282 miles; why not speak of a 'light-mile' as the time required for light (in a vacuum) to cover a distance of a mile and find it equivalent to 1/186,282 seconds?

Why not, indeed? It would enable us to use the speed of light in a vacuum, that universal constant, in measuring not only the vast distances encountered in astronomy, but also the ultra-short intervals of time dealt with in nuclear physics.

The only drawback is that 186,282 is such an uneven number. However, by a curious coincidence, the metric system comes to the rescue, since 186,282 miles is almost equal to 300,000 kilometres, so that a 'light-kilometre' is approximately equal to 1/300,000 seconds. Or, to put it another way, three and a third 'light-kilometres' equals a hundredth of a thousandth of a second.

To express still smaller units of time, it is only necessary to consider light to be covering smaller and smaller distances.

We can begin by pointing out that a kilometre (100,000 centimetres) is equal to a million millimetres; that one millimetre (a tenth of a centimetre) is equal to a million millimicrons; and that one millimicron (a tenth of a millionth of a centimetre) is equal to a million fermis. The fermi, named in honour of Enrico Fermi, is equal to a tenth of a trillionth of a centimetre. Well, then:

173

1 'light-kilometre' = 1 million 'light-millimetres'
1 'light-millimetre' = 1 million 'light-millimicrons'
1 'light-millimicron' = 1 million 'light-fermis'

Now we can relate these units to conventional units of time:

3 1/3 'light kilometres' is a hundred thousandth of a second.
3 1/3 'light-millimetres' is a hundredth of a thousandth of a millionth of a second.
3 1/3 'light-millimicrons' is a hundredth of a thousandth of a trillionth of a second.
3 1/3 'light-fermis' is a hundredth of a thousandth of a millionth of a trillionth of a second.

Is there any point in going further down the scale? Probably not, at least now. One fermi is approximately the diameter of sub-atomic particles. A 'light-fermi' is therefore about the time for a ray of light to travel from one end of an atomic nucleus to the other. It is the time required for the fastest known motion to cover the smallest known tangible distance. Until the day comes when we discover something faster than the speed of light, or something smaller than sub-atomic particles, we are not likely ever to have to deal with a time interval smaller than the 'light-fermi.' As of now, it is the ultimate split of the second.

Of what value is this new type of unit?

Well, to say that the half-life of one type of particle is of the order of a hundredth of a millionth of a second, and of another of a thousandth of a trillionth of a second, leaves very little impression on the mind. These measurements fade into the common denominator of 'the unimaginably small.'

But say instead that the half-life of one type of particle is of the order of a 'light-metre,' and the other of a 'tenth of a light-micrometre,' and visualization is easy. We are used to thinking in distances, and the difference between a metre and a tenth of micrometre is a vivid one to our consciousness.

Furthermore, sub-atomic particles often move at speeds near that of light. A particle with a half-life of about 30 'light millimetres' will have a chance to leave a bubble-chamber track 30 millimetres long before breaking down.

The one implies the other. By using conventional units, you might say that a length of track of about 30 millimetres implies a half-life of about a trillionth of a second (or vice versa), but there is no obvious connection between the two numerical values. To say that a track of 30 millimetres implies a half-life of 30 'light-millimetres' would be equally true and how neatly the two would fit together!

9
War

A Delay in the Taking of Syracuse

Archimedes of Syracuse, born around 287 B.C., was probably the greatest scientist of classical times. He performed legendary feats in mechanics ('give me a lever and a place to stand on, and I will move the world'), he was the first to calculate a reasonably accurate value of pi, the ratio between the circumference of a circle and its diameter (3.142 . . . etc.) and he estimated how many grains of sand would fill the universe — assuming, of course, that one knew the size of the universe.

In one of the best-known stories of his life, he was asked by his king and patron, Hieron II, to investigate an alleged crime. The anecdote is told by Vitruvius Pollio:

Hieron asked Archimedes to discover, without damaging it, whether a certain crown or wreath was made of pure gold, or if the goldsmith had fraudulently alloyed it with some baser metal.

While Archimedes was turning the problem over in his mind, he chanced to be in the bath house. There, as he was sitting in the bath, he noticed that the amount of water that was flowing over the top of it was equal in volume to that part of his body that was immersed. He saw at once a way of solving the problem. He did not delay, but in his joy leaped out of the bath. Rushing naked through the streets towards his home, he cried out in a loud voice that he had found what he sought. For, as he ran, he repeatedly shouted in Greek: *'Eureka! Eureka!* I've found it! I've found it!'

On his return home, Archimedes did the same test on the 'gold' crown as he had on his own person. It turned out to be partly made of silver, and the unlucky goldsmith was executed.

Archimedes turned from detective to military hero when, in 214 BC, the Roman general Marcellus besieged Syracuse by sea and land. As the historian Plutarch relates in this exciting narrative, Archimedes kept the Roman fleet at bay for three years with his lethal mechanical inventions:

When the Romans attacked them, the people of Syracuse were struck dumb with terror, imagining that they could not possibly resist such numerous forces and so furious an assault. But Archimedes soon began to ply his engines, and they shot against the Roman land forces all sorts of missiles and enormous stones, with so incredible a noise and rapidity that nothing could stand before them. They overturned and crushed whatever came in their way, and spread terrible disorder throughout the ranks.

On the side of the city towards the sea were erected vast machines, putting forth on a sudden, over the walls, huge beams with the necessary tackle, which, striking with a prodigious force on the Roman galleys, sank them at once: while other ships, hoisted up at the prows by iron grapples or hooks, like the beaks of cranes, and then dropped stern first, were plunged to the bottom of the sea. Others were pulled towards the shore by ropes and grapples, and after being whirled about and dashed against the rocks that projected below the walls, were broken to pieces and the crews perished.

Very often a ship lifted high above the sea, suspended and twirling in the air, presented a most dreadful spectacle. There it swung till the men were

thrown out by the violence of the motion, and then it split against the walls or sunk, on the engine's letting go its hold.

As for the engine which Marcellus brought forward upon eight galleys, called *Sambuca*, on account of its likeness to the musical instrument of that name,* while it was at a considerable distance from the walls, Archimedes discharged a stone at it of twenty-five pounds weight, and after that a second and a third, all of which striking upon it with an amazing noise and force, shattered and totally disjointed it.

Marcellus, in this distress, drew off his galleys as fast as possible, and sent orders to the land forces to retreat likewise. He then called a council of war, in which it was resolved to come close to the walls, if it was possible, the next morning before day. For the engines of Archimedes, they thought, being very strong, would then discharge themselves over their heads. But for this Archimedes had long been prepared, having by him engines fitted for all distances. Besides, he had caused holes to be made in the walls, in which he placed *scorpions*,** that did not carry very far, but could be discharged very fast.

When, therefore, the Romans got close to the walls, undiscovered as they thought, they were welcomed by a shower of darts and huge pieces of rock, which fell, as it were perpendicularly on their heads: for the engines played from every quarter of the walls. This obliged them to retire: and when they were at some distance, other shafts were shot at them, in their retreat, from the larger machines, which made terrible havoc among them as well as greatly damaging their shipping, without any possibility of their annoying the Syracusians in their turn. For Archimedes had placed most of his engines under cover of the walls; so that the Romans being infinitely distressed by an invisible enemy, seemed to fight against the gods.

But Marcellus laughed at his own artillery-men and engineers. 'Why do we not leave off contending with this mathematical Briareus?'*** said he, 'who, sitting on the shore and acting, as it were, in jest, has shamefully baffled our naval assault; and in striking us with a hundred bolts at once, exceeds even the hundred-handed giants in the fable?'

And in truth all the rest of the Syracusians were no more than a body in the battery of Archimedes, while he himself was the commanding brain. All other weapons lay idle. His were the only offensive and defensive arms of the city. At last the Romans were so terrified, that if they saw but a rope or a stick put over the walls, they cried out that Archimedes was levelling some machine at them, and turned their backs and fled. Marcellus, seeing

---

*   A stringed instrument, so presumably the weapon was some kind of catapult.

**   A massive mechanical bow. See a drawing of this vicious-looking weapon in Vol. 8, Page 400, of the 1967 *Encyclopedia Britannica*.

*** Briareus, a legendary Greek god of the seas, a giant with a hundred arms and ferocious strength.

this, gave up all thoughts of proceeding by assault, and leaving the matter to time, turned the siege into a blockade.

Yet Archimedes had such a depth of understanding, such a dignity of sentiment, and so copious a fund of mathematical knowledge, that, though in the invention of these machines he gained the reputation of a man endowed with divine rather than human knowledge, yet he did not vouchsafe to leave any account of them in writing. For he considered all attention to *mechanics*, and every art that ministers to common uses, as mean and sordid, and placed his whole delight in those intellectual speculations which, without any relation to the necessities of life, have an intrinsic excellence arising from truth and demonstration only.*

But we are not to reject as incredible what is related of him, that being perpetually charmed by a siren, that is his geometry, he neglected his meat and drink, and took no care of his person: that he was often carried by force to the baths, and when there, he would make mathematical figures in the ashes, and with his finger draw lines upon his body when it was anointed; so much was he transported with intellectual delight, and such an enthusiasm in science; and though he was the author of many curious and excellent discoveries, yet he is said to have desired his friends only to place on his tombstone a cylinder containing a sphere, and to set down the proportion which the containing solid bears to the contained. Such was Archimedes, who exerted all his skill to defend himself and the town of Syracuse against the Romans.

After three years of blockade, during which the Roman fleet kept well out of range of any conceivable weapon that Archimedes might invent, Syracuse fell. Plutarch describes the death of the great military scientist:

What most affected Marcellus was the unhappy fate of Archimedes; who was at that time in his study, engaged in some mathematical researches; and his mind was so intent upon his diagram that he neither heard the tumultuous noise of the Romans nor perceived that the city was taken.

A soldier suddenly entered his room, and ordered him to follow him to Marcellus: but Archimedes refused to do it until he had finished his problem.** The soldier, in a passion, drew his sword and killed him. Marcellus was much concerned at his death. He turned away his face from his murderer, as from an impious and execrable person: and having by

---

* Thus Archimedes, like most scientists of ancient Greece and Rome, was under the malign influence of Socrates and Plato, who held that the proper function of science was to sharpen the wits and exercise the mind, and that its *applications* were to be despised. This crippling philosophy held back the progress of mankind for more than a thousand years.

** 'Don't disturb my circles!' he is said to have shouted imperiously.

enquiry found the relations of Archimedes, he bestowed upon them many signal favours.

It might have imagined that the Romans would have profited from the lesson taught them by Archimedes, ever to seek ways to improve their military technology. But they had no incentive to do so. After their defeat of Hannibal in the Second Punic War a few years later, they faced no formidable enemies for many centuries. Eventually, in consequence of the lethargy of this long peace, the Empire fell, partly to the superior military science of the barbarians.

Roman martial skills had much declined by the end of the fourth century AD; but even at the peak of their strength they might have had difficulty facing the mounted archers of the Huns, who shot their heavy arrows at full gallop, and in the words of Edward Gibbon, 'with unerring aim and irresistible force.' Archimedes would have appreciated the Huns.

'Those Wondrous New-made Bombards'

Cannon, first introduced to European battlefields in the fourteenth century, utterly revolutionized warfare. The only exciting account of this invention I was able to find is from Sir Arthur Conan Doyle's historical novel *Sir Nigel*. The events described below take place in about 1350, during the Hundred Years' War between England and France.

It was a bright sunshiny morning when Nigel Loring found himself at last able to leave his turret chamber and to walk upon the rampart of the castle. His eyes finally settled upon a strange object at his feet.

It was a long trumpet-shaped engine of leather and iron bolted into a rude wooden stand and fitted with wheels. Beside it lay a heap of metal slugs and lumps of stone. The end of the machine was raised and pointed over the battlement. Behind it stood an iron box which Nigel opened. It was filled with a black coarse powder, like gritty charcoal.

'By Saint Paul!' said he, passing his hands over the engine. 'I have heard men talk of these things, but never before have I seen one. It is none other than one of those wondrous new-made bombards.'

'In sooth it is even as you say,' answered Aylward the archer, looking at it with contempt and dislike. 'I have seen them here upon the ramparts, and have also exchanged a buffet or two with him who had charge of them. He was jack-fool enough to think that with this leather pipe he could outshoot the best archer in Christendom. I lent him a cuff on the ear that laid him across his foolish engine.'

'It is a fearsome thing,' said Nigel, who had stopped to examine it. 'We lived in strange times when such things can be made. It is loosed by fire, is it not, which springs from the black dust?'

'By my hilt! fair sir, I know not. And yet I call to mind that ere we fell out this foolish bombardman did say something of the matter. The fire-dust is within and so also is the ball. Then you take more dust from this iron box and place it in the hole at the further end — so. It is now ready. I have never seen one fired, but I wot that this one could be fired now.'

'It makes a strange sound, archer, does it not?' said Nigel, wistfully.

'So I have heard, fair sir — even as the bow twangs, so it also has a sound when you loose it.'

'There is no one to hear, since we are alone upon the rampart, nor can it do scathe since it points to sea. I pray you to loose it and I will listen to the sound.'

Nigel bent over the bombard with an attentive ear, while Aylward, stooping over the touch-hold, scraped away diligently with a flint and steel. A moment later both he and Nigel were seated some distance off upon the ground, while amid the roar of the discharge and the thick cloud of smoke they had a vision of the long black snake-like engine shooting back on the recoil. For a minute or more they were struck motionless with astonishment, while the reverberations died away and the smoke-wreaths curled slowly up to the blue heavens.

'Good lack!' cried Nigel at last, picking himself up and looking around him. 'Good lack, and Heaven be my aid. I thank the Virgin that all stands as it did before. I thought that the castle had fallen.'

'Such a bull's bellow I have never heard,' cried Aylward, rubbing his injured limbs. 'I would not touch one again — not for a hide of the best land in Puttenham!'

'It may fare ill with your own hide, archer, if you do,' said an angry voice behind them. Sir John Chandos had stepped from the open door of the corner turret and stood looking at them with a harsh gaze. Presently, as the matter was made clear to him, his face relaxed into a smile.

'Hasten to the warder, archer, and tell him how it befell. You will have the castle and the town in arms. And you, Nigel, how in the name of the saints came you to play the child like this?'

'I knew not its power, fair Lord.'

'But my soul, Nigel, I think that none of us know its power. I can see the day when all that we delight in, the splendour and glory of war, may all go down before that which beats through the plate of steel as easily as the leathern jacket. I have bestrode my war-horse in my armour and have looked down at the sooty, smoky bombardman beside me, and I have thought that perhaps I was the last of the old and he the first of the new; that there would come a time when he and his engines would sweep you and me and the rest of us from the field.'

Einstein and the Bomb

Few discoveries have had such momentous effect on mankind as Albert Einstein's Special Theory of Relativity, which was later to give us the atomic and hydrogen bombs as well as peaceful nuclear energy.

This is the story of the coming of the atomic bomb. It took forty years to happen, from Einstein's first glimpses of the theory in 1905 while he strolled through the parks of Berne, to the explosion that shattered the dawn of New Mexico in 1945. Special Relativity has many aspects, but the one that concerns us here is his famous equation $E=MC^2$, which shows how, at sufficiently high temperatures — such as are found at the core of the Sun — matter is converted to energy. Now, the 'C' in the equation denotes the speed of light, which is a fantastic 186,282 miles per second. It will thus be seen that a very small amount of matter (in the form of uranium or plutonium) will release a gigantic amount of energy.* This is the principle of atomic power, although in 1905 not even Einstein believed that man would ever be able to exploit it.

Einstein's most readable biographer Peter Michelmore captures the atmosphere of Berne in 1905, where the dreamy young scientist worked as a clerk at the Swiss Patent Office, working out his theories on scraps of paper when his employer was not looking:

---

* At the core of the Sun, for example, where the temperature is 30 million degrees C, some 4 million tons of hydrogen are being converted to helium and to energy every second by thermonuclear fusion.

181

Einstein's paper in the journal *Annalen de Physik** was noted by his academic acquaintances in Berne, but they were not prepared to accept the word of a young patent officer as gospel on such momentous matters. In the after-work coffee house discussions, it was Einstein versus the Rest. They argued particularly about his simple statement that $E=MC^2$.

'You're saying there's more horsepower in a lump of coal than in the whole Prussian cavalry,' they complained. 'If this were true, why hasn't it been noticed before?'

'If a man who is fabulously rich never spent or gave away a penny,' Einstein replied, 'then no one could tell how rich he was, or even whether he had any money at all. It is the same with matter. So long as none of the energy is given off externally, it cannot be observed.'

'And how do you propose to release all this hidden energy?'

'There is not the slightest indication that the energy will ever be obtainable,' said Einstein. 'It would mean that the atom would have to be shattered at will. There is scarcely a sign that this will be possible. We see atomic disintegration only in nature, as in the case of radium. Radium's activity depends on the continual explosive decomposition of the atom.'

The others wanted to know how he had worked out his energy equation from radium experiments. To their horror, Einstein said he had not been inside a laboratory for years.

---

*   It was entitled 'On the Electrodynamics of Moving Bodies.' The full text may be found in Einstein's co-authored book *The Principle of Relativity*. See the Bibliography for details.

'Then this Relativity of yours is all fiction,' they said. 'It is something you dreamed up. A proper physicist makes his discoveries by fresh experimentation and checking his results. It's the only way.'

'Rubbish,' said Einstein, the twenty-six-year-old civil servant. 'Physics is a logical system of thought in a state of evolution. Its basis cannot be obtained merely by experiment and experience. Its progress depends on free invention.' He added, however, that Relativity must stand the test of human experience before it could be fully accepted. He smiled. 'I haven't the faintest doubt that I am right.'

His doubts that the atom could ever produce power which man could tap did not last long. The 1920s and 1930s saw tremendous progress in atomic discovery. It was soon discovered that huge temperatures were not necessary to release the energies locked up in matter. This could be done by bombarding atoms with other atoms. Lord Rutherford had established the basis of atomic theory by his revelation that atoms consist of a loose structure of electrons encircling a heavy central core, the nucleus. In 1919 he succeeded in splitting the hydrogen atom, becoming the first to achieve an artificial nuclear reaction.* In 1932 Sir James Chadwick took this process further by bombarding atoms with neutron particles. By 1938 Otto Hahn and Lise Meitner had discovered the full principle of nuclear fission, or 'splitting' — although it was not until 1942 that Enrico Fermi achieved the first sustained fission reaction in a Chicago squash court.

But along with these advances came the rising menace of Hitler's power. Knowing that it was possible to build an atomic bomb, the atomic scientists (many of whom had fled from Nazi Germany) began to fear that the Germans would be first to build it, and in all probability use it. As the war began Einstein, who had become an American citizen, was living in semi-retirement under the assumed name of Dr Moore to avoid the attentions of journalists. His colleagues knew that the free nations must build the bomb before the Germans. But they also knew that only Einstein had enough prestige to persuade the American Government to start the necessary work. Two of them set out on a visit to Einstein.

Yet, as David Bergamini and Henry Margenau relate in their book *The Scientist*, the most difficult part of their task was *finding* him:

---

* Even as late as 1933, Rutherford could see no practical use for nuclear power. 'Anyone who looks for a source of power in the transformation of atoms is talking moonshine,' he told the British Association for the Advancement of Science.

One muggy day in July, 1939, two eminent scientists in an automobile found themselves lost in the wilds of Long Island. They had come on a mission so disconcertingly important, so melodramatic and irregular, that they had neglected to make sure of their directions.

'Perhaps I misunderstood on the telephone,' one of them ventured. 'I thought he said Patchogue.'

'Could it have been Cutchogue?' asked the other, a trifle irked.

Some time later they pulled up on a street in Cutchogue, and asked the way to 'Dr. Moore's cabin,' but to no avail. As they drove around, feeling more and more frustrated, one of them said:

'Maybe fate never intended this. Let's go home.'

'Wait,' suggested the other. 'What if we simply inquire where Einstein is staying?'

They stopped at the curb beside a small boy of about seven, and asked if he knew Professor Einstein.

'Sure,' he said. 'Want me to take you to him?'

And so it came about, the story goes, that Eugene Wigner and Leo Slizard, two Hungarian refugee physicists, finally reached Albert Einstein on an afternoon two months before the start of World War II and persuaded him to write a letter to President Roosevelt urging quick action to offset possible Nazi progress on building an atomic bomb.*

After much discussion, this is the famous letter that Einstein finally wrote to President Roosevelt:

Sir:

Some recent work by E. [Enrico] Fermi and L. Slizard, which has been communicated to me in manuscript, leads me to expect that the element uranium may be turned into a new and important source of energy in the immediate future. Certain aspects of the situation which has arisen seem to call for watchfulness and, if necessary, quick action on the part of the Administration. I believe therefore that it is my duty to bring to your attention the following facts and recommendations:

In the course of the last few months it has been made probable — through the work of Joliot [Frédéric Joliot-Curie] in France as well as Fermi and Slizard in America — that it may become possible to set up a nuclear

---

* Six years earlier, in October, 1933, Slizard had thought of a way to make nuclear weapons while crossing Southampton Row in London. 'As the light turned to green,' he wrote later, 'it suddenly occurred to me that if we could find a chemical element which is split by neutrons, and which would emit *two* neutrons when it absorbed *one* neutron, such an element, in sufficiently large mass, could sustain a nuclear chain reaction.' One wonders how history might have changed if the light had still been red earlier and the absent-minded Slizard had failed to notice.

chain reaction by which vast amounts of power and large quantities of radium-like elements would be generated. Now it appears almost certain that this could be achieved in the immediate future.

This new phenomenon would also lead to the construction of bombs, and it is conceivable — though much less certain — that extremely powerful bombs of a new type may thus be constructed. A single bomb of this type, carried by boat and exploded in a port, might very well destroy the whole port together with some of the surrounding territory. However, such bombs might very well prove to be too heavy to be carried by air.

The United States has only very poor ores of uranium in moderate quantities. There is some good ore in Canada and in former Czechoslovakia, while the most important source of uranium is the Belgian Congo.

In view of this situation you may think it desirable to have some permanent contact maintained between the Administration and the group of physicists working on chain reactions in America . . . [*Einstein then goes into details as to how this work might be co-ordinated between various Government departments.*] *He continues*:

I understand that Germany has actually stopped the sale of uranium from the Czechoslovakian mines which she has taken over. That she should have taken such early action might perhaps be understood on the ground that the son of the German Under-Secretary of State, von Weizsacker, is attached to the Kaiser-Wilhelm Institut in Berlin, where some of the American work on uranium is now being duplicated.

Yours very truly,

Albert Einstein

This letter was handed personally to President Roosevelt in the White House by his adviser Alexander Sachs, who was sympathetic to the physicists. Michelmore continues the story:

Roosevelt was impressed but not convinced. He reminded Sachs that the Navy had virtually rejected earlier bids for help from Fermi and Slizard. He did not see how he could go against his service chiefs. Sachs left his portfolio on Roosevelt's desk and asked for another appointment the next day. He was invited to breakfast. Einstein's letter had not been decisive after all, so Sachs dreamed up a different approach.

At breakfast Sachs launched into a long story of how Napoleon once scoffed at the idea of the young American inventor, Robert Fulton, to build steamships that would transport French invaders safely and quickly across the Channel to England.*

* 'What, sir?' the Emperor is reported to have snapped at Fulton after listening to him impatiently for a few moments. 'You would make a ship sail against the wind and currents by lighting a bonfire under decks? I pray you excuse me. I have no time to listen to such nonsense.'

Roosevelt eyed his adviser quizzically for a moment.

'Alex, what you're after is to see that the Nazis don't blow us up?' he asked.

'Precisely,' said Sachs.

From an adjoining room, Roosevelt called in General 'Pa' Watson, his military secretary, slapped down the sheaf of documents that included Einstein's letter and told him: 'This requires action.'

That was the beginning of what became the two billion dollar Manhattan Project to build the atomic bomb.

The Manhattan Project started in earnest in 1943. Dozens of atomic scientists, working under J. Robert Oppenheimer, laboured to build the bomb that was to explode just before dawn at Alamagordo, New Mexico, on 16 July 1945. Lancing Lamont, in his book *The Day of Trinity*, describes the explosion that changed the world for ever:

A pinprick of brilliant light punctured the darkness, spurting upward in a flaming jet, then bleached the desert to a ghastly white. It was 5:29:45 A.M.

Oppenheimer, in that blinding instant, thought of fragments from the sacred Hindu epic Bhagavad-Gita:

> *If the radiance of a thousand suns*
> *Were to burst at once into the sky*
> *That would be like the splendour of the Mighty One . . .*
> *I am become Death,*
> *The shatterer of worlds.*

For a fraction of a second, the light in that bell-shaped fire was greater than any ever produced on earth. It could have been seen from another planet. The temperature at its centre was four times that at the core of the sun and more than ten thousand times that at the sun's surface. The pressure, caving in the ground beneath, was over 100 billion atmospheres. The radioactivity emitted was equal to one million times the world's total radium supply.

No living thing touched by that raging furnace survived. Within a millisecond, the fireball had struck the ground, flattening out at its base and acquiring a skirt of molten black dust that boiled and billowed in all directions. Within twenty five milliseconds, the fireball had expanded upwards to a point where the Washington Monument would have been enveloped. At eight tenths of a second, the ball's white-hot dome would have topped the Empire State Building. The shock wave raced across the desert.

People turned to look at the fireball, inflated now half a mile wide, and wondered if it would ever stop growing. In their excitement, many threw off their dark glasses and instantly lost sight of what they had waited many years to see. At Base Camp there were silent handclasps and murmurs of amazement. A squabble of excited voices rose to a deafening din. A piercing whoop was followed by a mad jig that suggested to one observer the rites of prehistoric savages.

The intercom came alive with a babble of voices as scientists at Base Camp and other shelters filled the air waves with congratulatory cries. Someone started a snake dance, and the happy line curled around the control room. Asleep in an empty barrack, a soldier who had gone on a beer binge the night before felt nothing until the blast jolted him from his stupor. He opened his eyes to the north and screamed in pain.

Enrico Fermi didn't notice the crack of the shock wave which had sounded across the desert like a thunderclap. He was too engrossed in dribbling scraps of paper from his pockets. He watched them slowly fall, then sweep suddenly away as the shock wave struck them. Within seconds, Fermi had calculated the distances the scraps had blown and estimated the force of the explosion as equivalent to 20,000 tons of TNT.

When the shock wave reached his trench, the first reaction of the chief of security, General Groves, was to announce: 'We must keep this whole thing quiet.' But his deputy Major Stephens leaned over and said: 'Sir, I think they heard the noise in five states.'

Postscript

The Nazis never came near to building the bomb, despite the scientists' fears. Hitler lost his early interest in the project, believing it would take too long to influence the outcome of the war. Himmler actively sabotaged it with his habit of arresting Germany's atomic scientists, either for suspected disloyalty to the State or else because he wanted them to work on his pet pseudo-scientific schemes.

One month after the Alamagordo explosion, two American atomic bombs were dropped on Japanese cities, killing more than 100,000 people, and bringing about the end of the Second World War. Since this terrible example of what nuclear weapons can do, the world has arguably been a safer place. And, as Jacob Bronowski later remarked, 'science has nothing to be ashamed of, even in the ruins of Nagasaki'.

'Expert' Opinion

Scientific 'experts' are fond of assuring us of the futility of the American 'Star Wars' project, to destroy incoming enemy ballistic missiles before they can reach their targets.

A similar technical assurance was given to a Congressional committee in 1945 by the electronics pioneer Dr Vannevar Bush — about ballistic missiles themselves:

People have been writing about a 3,000-mile high rocket shot from one continent to another, carrying an atomic bomb which would land exactly on a certain target, such as a city.

I say, technically, that I don't think anyone in the world knows how to do such a thing, and I feel confident that it will not be done for a very long period of time to come. I think we can leave that out of our thinking. I wish the American people would leave that out of their thinking.

Five years after he spoke, Russia developed the first intercontinental ballistic missile.

10
Science Fiction
Why It Comes True

The phrase 'the wildest science fiction' is generally held synony-
mous with fantastical nonsense. The science-fiction writer Ben Bova
challenges this view.

It is called FORECASTS. It was created for the Joint Chiefs of Staff, the
generals and admirals who head the U.S. Army, Navy, Air Force and Marine
Corps. It has cost more than a million dollars to develop, and it will cost
still more before it is fully tested and operational.

FORECASTS is a computer model of the whole world. It is a highly
complex computer program with enormous amounts of data about global
political trends, natural resources, and social and economic factors. The
Joint Chiefs will use FORECASTS to help them make the predictions that
go into their Joint Long Range Strategic Appraisal, in which the JCS
evaluate what the world in general, and certain nations in particular, will
look like over the next thirty years.

Science fiction writers have been making such predictions for generations now, and because the accuracy of the forecast is only as good as the
quality of the information being used, the predictions of science fiction
writers have generally been better than those of anyone else's — including
the complex computerized 'world models' of the scientists who call
themselves futurists.

Futurists like the late Herman Kahn, for example, have consistently
missed the major turning points in recent history. No futurist predicted the
Arab oil embargo of 1973 and the resulting energy crisis. The Club of
Rome's much-heralded study of 1971, *The Limits to Growth*, failed utterly
to understand that the Earth is not the only body in the universe from
which energy and natural resources can be extracted. The Presidential
Commission of 1979 which produced *Report on the Year 2000* was equally
medieval in its view, and failed even to see the vigorous growth of living
standards in the small industrializing nations of the Far East, nations such
as Taiwan, South Korea, Malaysia and Singapore.

Science fiction's record of predicting the future is much better. Atomic power, space flight, organ transplants, population explosions, genetic engineering — all these changes in human capabilities were described in science fiction stories at least 30 years before they took place. What is more, science fiction writers also predicted the social consequences of such changes: the Cold War stalemate that resulted from atomic weapons; the urban sprawl that came from increased mobility and growing population; the breakdown of traditional family values that has accompanied the new sexual freedom.

Why is it that science fiction writers have seen further into the future than all others — and more clearly? Is it because they are trained in the sciences? Hardly. Although many writers of science fiction have degrees in the physical or social sciences, very few of them are actually practising scientists. Isaac Asimov, for instance, has not engaged in scientific research for nearly three decades, despite his doctorate in chemistry and his title of professor of biochemistry at Boston University School of Medicine. Ray Bradbury, on the other hand, has no scientific training at all. Yet both Asimov and Bradbury are world-class science fiction writers, and both have produced scores of powerful and predictive stories.

The thing that makes a science fiction writer better at predicting the future than anyone else is not scientific knowledge (although an understanding of science is very helpful, even necessary.) Nor is it a mystical, arcane, extrasensory perception of the future.

His or her secret can be told in two words: *freedom* and *imagination*.

The professional scientists who try to predict the future with computerized accuracy always fail because *they are required to stick to the facts*. No futurist is going to predict that a semi-accidental discovery will transform the entire world. Yet the invention of the transistor did just that: without the transistor and its microchip descendants, today's world of computers and communication satellites simply would not exist. Yet a futurist's forecast of improvements in electronic technology, made around 1950, would have concentrated on bigger and more complicated vacuum tubes and missed entirely the microminiaturization that transistors have made possible. Science fiction writers, circa 1950, 'predicted' marvels such as wrist-radios and pocket-sized computers, not because they foresaw the invention of the transistor, but because they instinctively felt that some kind of improvement would come along to shrink the bulky computers and radios of that day.*

The professional futurists labour under this enormous handicap; they are not allowed to consider the 'wild cards,' the crazy things that can and usually do happen. They are restricted to more or less straight-line

---

* The Moon-landing missions did more, perhaps, than anything else to shrink the size of computers. — A.B.

extrapolation of the facts as we know them today. But science fiction writers have the freedom to use more than the facts. They can use their imaginations. They can ask themselves: 'What would happen if . . .' and then set out to write a story that answers that question. They can use their knowledge of the human soul — for that is what fiction is all about — not merely to describe the marvellous invention or the strange discovery, but to betray how real people — you or I — might react to these new things.

That is science fiction's great advantage, the freedom to employ imagination to its fullest. The science fiction writer is not required to be accurate, merely entertaining. Although the writer need not have a professional knowledge of science, he or she should understand the basics well enough to know what is impossible — and how to move at least one step beyond that limit. The rule of thumb in good science fiction is that you are free to invent anything you like, providing no one else can *prove* that it could never be. Even though physicists are certain that nothing in the universe travels faster than the speed of light, they cannot *prove* that it is utterly impossible for a starship to circumvent that speed limit; therefore science fiction writers can create interstellar dramas, with merely a slight bow to acknowledge that their faster-than-light starships are using principles unknown in the 20th century. In creating such stories, about some future times and places, the writer often creates an inner reality that eventually comes true.

You don't need a million-dollar computer program or a team of Pentagon scientists. All you need is talent, plus the fortitude to work long and lonely hours, together with the freedom to let your imagination roam where it will.

The New Atlantis

Francis Bacon was much more than the courtier who remarked cynically that 'all rising to great place is by a winding stair'. He was a 'philosopher of science.' He saw more deeply into the future than any person of his age — and perhaps of any age.

'Knowledge is power.' 'I have taken all knowledge to be my province.' 'The age of the centuries is the youth of the world.' 'Among all the benefits that could be conferred among mankind, I found none so great as the discovery of new arts, endowments and commodities for the betterment of man's life.' These are among his most famous aphorisms.

In 1627 was published his posthumous work *The New Atlantis*, one of the first science-fiction novels. It was a best-seller for more than a century. Its vision of the future was extraordinary. His heroes are shipwrecked in some remote region of the Pacific, where they discover the famous republic of Bensalem. Its citizens are enormously advanced. They have aircraft and submarines, refrigerators and hearing aids. They are obsessed with 'profitable inventions', and what they cannot invent themselves they steal from others.

In this extract from the novel their leader explains to his visitors what happens in Solomon's House, the scientific heart of Bensalem:

God bless thee, my son. I will impart unto thee a relation of the true state of Solomon's House, and I will keep this order. First, I will set forth the end [aim] of our foundation. Second, the preparations and instruments we have for our works. Third, the several employments and functions whereto our fellows are assigned. And fourth, the ordinances and rites which we observe.

The End of our Foundation is the knowledge of Causes, and secret motions of things; and the enlarging of the bounds of human empire, to the effecting of all things possible.

The Preparations and Instruments are these. We have large and deep caves of several depths: the deepest are sunk 600 fathoms: and some of them are digged and made under great hills and mountains: so that if you reckon together the depth of the hill and the depth of the cave, some of them are above three miles deep. And we use these caves them, for all indurations,* refrigeration, and natural mines; and producing new artificial metals. We use them also for curing diseases, and for prolongation of life in those that choose to live there, who are well accommodated of all things necessary; and indeed live very long; by whom we learn many things.

We have burials in several earths, where we put diverse cements, as the Chinese do their porcelain. But we have also great variety of composts, and soils, to make the earth fruitful.

We have high towers; the highest about half a mile in height; and some of them set upon high mountains: so that the vantage of the hill with the tower is in the highest of them three miles at least. We use these towers, according to their several heights and situations, for insulation, refrigeration, conservation; and for the view of winds, rain, snow, hail; and some of the fiery meteors also. And upon them, in some places, are dwellings of hermits, whom we instruct what to observe.

We have great lakes, both salt and fresh, that we use for fish and fowl, and for burials of some natural bodies: for we find a difference in things buried in earth or in air below the earth, and things buried in water. In

* Hardenings

some pools, we strain fresh water into salt. We have some rocks in the midst of the sea, and some bays upon the shore, for some works that require the air and vapour of the sea. We have likewise violent streams and cataracts, which serve us for many motions; and engines to multiply and enforce the winds.

We have also a number of artificial wells and fountains, made in imitation of the natural sources and baths; as tincted upon vitriol, sulphur, steel, brass, lead, nitre and other minerals. And again we have little wells for infusions of many things, where the waters take the virtue quicker and better than in vessels or basins. And amongst them we have a water which we call Water of Paradise, being, by that we do to it, made very sovereign for health and prolongation of life.

We have fair and large baths, of several mixtures, for the cures of diseases, and the restoring of man's body from arefaction,* and others for the confirming of it in strength of sinews, vital parts, and the very juice and substance of the body.

We have also large and various orchards and gardens, wherein we do not so much respect beauty, as variety of ground and soil, proper for divers trees and herbs; and some very spacious, where trees and berries are set whereof we make divers kinds of drinks, besides the vineyards. In these we practise grafting and inoculating, as well of wild-trees as fruit-trees, which produceth many effects. And we make, by our art, trees and flowers to come earlier or later than their seasons; and to bear fruit more speedily than by their natural course. We make them also by art greater much than their nature; and their fruit greater and sweeter and of differing taste, smell and colour from their nature.

We have means of making plants rise without seeds; and making new plants, differing from the vulgar; and of making one plant turn into another.

We have also parks for all sorts of beasts and birds, which we use not only for view of rareness, but for dissections and trials; that thereby we may take light what may be wrought upon the body of men. Wherein we find many strange effects; resuscitating some that seem dead in appearance and the like. We try also all poisons and other medicines upon them. By art likewise, we make them greater or taller than their kind is; and contrary-wise barren and not generative. Also we make them differ in colour, shape and activity. We find means to make commixtures and copulations of different kinds; which have produced many new kinds. We make several kinds of serpents, worms, flies, fishes from putrefaction;** whereof some are advanced to be perfect creatures, like beasts or birds; and have sexes,

---

*   Decomposition
** It was not until the nineteenth century that Louis Pasteur completed the eighteenth century work of Lazzano Spallanzani in disproving the old theory of 'spontaneous generation' — that lower animals were created from putrefaction.

and do propagate. Neither do we this by chance, but we know beforehand from what matter and commixture those creatures will arise.

We have divers mechanical arts which you have not, and stuffs made by them; as papers, linen, silks, tissues, dainty works of feathers of wonderful lustre; excellent dyes, and many others and shops likewise.

We have also furnaces of great diversities of heats: fierce and quick; strong and constant; soft and mild; blown, quiet, dry, moist and the like. But above all we have heats, in imitation of the sun's heat, whereby we produce admirable effects. Besides, we have heats of dungs; and of bellies and maws of living creatures, and of their bloods and bodies; and of hays and herbs laid up moist; of lime unquenched; and such like. Instruments also which generate heat only by motion. And farther, places for strong insulations; and again, places under the earth, which by nature or art yield heat.

We make demonstrations of all lights, and radiations; and of all colours; and out of things uncoloured and transparent, we can represent unto you all several colours; not in rainbows, as it is in gems and prisms, but of themselves single.

We procure means of seeing objects afar off; as in the heaven and remote places; and represent things near as afar off; and things afar off as near; making feigned distances. We have also helps for the sight, far above spectacles and glasses that are in use. We have also glasses and means to see small and minute bodies perfectly and distinctly; as the shapes and colours of small flies and worms, grains and flaws in gems which cannot otherwise be seen; observations in urine and blood, not otherwise to be seen. We make artificial rainbows, halos, and circles about light. We represent also all manner of reflexions, refractions, and multiplications of visual beams of objects.

We have also precious stones of all kinds, many of them of great beauty, and to you unknown; crystals likewise; and glasses of divers kinds; and amongst them some of metals vitrified, and other materials besides those of which you make glass. Also a number of fossils, and imperfect minerals, which you have not. Likewise loadstones of prodigious virtue; and other rare stones, both natural and artificial.

We have sound-houses, where we practise and demonstrate all sounds, and their generation. We have harmonies which you have not; divers instruments of music likewise to you unknown, some sweeter than any you have; together with bells and rings that are dainty and sweet. We represent small sounds as great and deep; likewise great sounds extenuate and sharp. We imitate all articulate sounds, and the voices and notes of beasts and birds. We have certain helps which, set to the ear, do further the hearing greatly. We have also divers strange and artificial echoes, reflecting the voice many times and as it were tossing it; and some that give back the voice louder than it came; some shriller, and some deeper; yet some rendering the voice differing in the letters or articulate sound from that they receive. We have also means to convey sounds in trunks and pipes, in strange lines and distances.

We have engine-houses, where are prepared engines and instruments for all sorts of motions. With these we make swifter motions than any you have, either out of your muskets or any of your engines; and to make them and multiply them more easily, and with small force, by wheels and other means: and to make them stronger, and more violent than yours are; exceeding your greatest cannons and basilisks.* We represent also ordnance and instruments of war, and engines of all kinds; and new mixtures and compositions of gunpowder. Also fireworks of all variety both for pleasure and use. We imitate also flights of birds; we have some degrees of flying in the air; we have boats for going under water, and brooking of seas;** also swimming-girdles and supporters. We have divers curious clocks; and other like motions of return, and some perpetual motions. We imitate also motions of living creatures, by images of men, beasts, birds, fishes, and serpents. We have also a great number of other various motions, strange for equality, fineness, and subtlety.

We have also a mathematical house, where are represented all instruments, as well of geometry as astronomy, exquisitely made.

We have also houses of deceits of the senses; where we represent all manner of feats of juggling, false apparitions, impostures, and illusions; and their fallacies. And surely you will easily believe that we, that have so many things truly natural which induce admiration, could in a world of particulars deceive the senses, if we would disguise those things and labour to make them seem more miraculous.

These are the riches of Solomon's House.

For the several employments and offices of our fellows; we have twelve that sail into foreign countries, under the names of those of other nations, (for our own we conceal;) who bring us the books, and abstracts, and patterns of experiments of all other parts. These we call Merchants of Light.***

We have three that collect the experiments which are in all books. These we call Depredators.

---

* Basilisk, a large brass gun of Tudor times. Named after the fabulous king of serpents.
   It is a basilisk unto mine eye,
   Kills me to look on't.
   — *Cymbeline*, Act 2, Scene 4.
** Taming the sea, turning the tempestuous ocean into the semblance of navigable rivers.
*** Today we would call them industrial spies.

We have three that collect the experiments of all mechanical arts; and also of liberal sciences; and also of practises which are not brought into arts. These we call Mystery-men.

We have three that try new experiments, such as themselves think good. These we call Pioneers of Miners.

We have three that draw the experiments of the former four into titles and tables, to give the better light for the drawing of observations and axioms out of them. These we call Compilers.

We have three that bend themselves, looking into the experiments of their fellows, and cast about how to draw out of them things of use and practise for man's life, and knowledge as well for works as for plain demonstration of causes, means of natural divinations, and the easy and clear discovery of the virtues and parts of bodies. These we call Dowry-men or Benefactors.

Then, after divers meetings and consults of our whole number, to consider of the former labours and collections, we have three that take care, out of them, to direct new experiments, of a higher light, more penetrating into nature than the former. These we call Lamps.

We have three others that do execute the experiments so directed, and report them. These we call Inoculators.

Lastly, we have three that raise the former discoveries by experiments into greater observations, axioms, and aphorisms. These we call Interpreters of Nature.

We have also, as you must think, novices and apprentices, that the succession of the former employed men do not fail; besides a great number of servants and attendants, men and women. And this we do also: we have consultations, which of the inventions and experiences which we have discovered shall be published, and which not: and take all an oath of secrecy, for the concealing of those which we think fit to keep secret: though some of those we do reveal sometimes to the State, and some not.

For our ordinances and rites: we have two very long and fair galleries: in one of these we place patterns and samples of all manner of the more rare and excellent inventions: in the other we place the statues of all principal inventors. There we have the statue of your Columbus that discovered the West Indies: also the inventor of ships: your monk that was the inventor of ordnance and of gunpowder:* the inventor of music: the inventor of letters; or printing: of observations of astronomy: of works in metal: of glass: of silk of the worm: of wine: of corn and bread: and of sugars. For upon every invention of value we erect a statue to the inventor to give him an honourable reward. These statues are some of brass; some of marble and touchstone; some of cedar and other special woods gilt and adorned: some of iron; some of silver; some of gold.

---

* Bacon's thirteenth century namesake Roger Bacon.

We have certain hymns and services, which we say daily, of laud and thanks to God for his marvellous works: and forms of prayers, imploring his aid and blessing for the illumination of our labours, and the turning of them into good and holy uses.

Lastly, we have circuits or visits of divers principal cities of the kingdom; where, as it cometh to pass, we do publish such new profitable inventions as we think good. And we do also declare natural divinations of diseases, plagues, swarms of hurtful creatures, scarcity, tempests, earthquakes, great inundations, comets, temperature of the year, and divers other things; and we give counsel thereupon what the people shall do for the prevention and remedy of them.

11
Diverse Matters
Bacon's Revolution

The second half of the seventeenth century saw the real birth of science, after a generation of tumult and civil war. The historian Macaulay describes this intellectual explosion:

The torrent which had been dammed up in one channel rushed violently into another. The revolutionary spirit, ceasing to operate in politics, began to exert itself in every department of science. The year 1660, that of the Restoration, saw also the foundation of the Royal Society, destined to be the chief agent in a long series of glorious reforms.

Experimental science became all the mode. The transfusion of blood, the study of air, the fixation of mercury, succeeded to that place in the public mind lately occupied by the disputes of the political clubs. Dreams of perfect forms of government made way for dreams of wings with which men were to fly from the Tower to the Abbey, and of double-keeled ships that would not founder in the fiercest storm.

All classes were hurried along by the prevailing sentiment. Cavalier and Roundhead, Churchman and Puritan, were for once allied. Poets sang of the approach of the golden age. Dryden, with more zeal than knowledge, foretold things which neither he nor anyone else understood. The Royal Society, he predicted, would lead us to the extreme edge of the globe, and there delight us with a better view of the Moon. Under the direction of Lord Keeper Guildford, the first barometers were sold in London. Prince Rupert has the credit of inventing mezzotint engraving. The King himself had a laboratory at Whitehall, and was far more attentive to it than to the Council board. It was almost necessary to the character of a fine gentleman to have something to say about air pumps and telescopes; and even fine ladies thought it becoming to affect a taste for science, breaking forth into cries of delight at finding that a magnet really attracted a needle, and that a microscope made a fly look as large as a sparrow.

The great work of interpreting nature was performed by the English as never before in any age by any nation. The spirit of Francis Bacon was abroad, a spirit admirably compounded of audacity and sobriety. There was a strong persuasion that the world was full of secrets of high moment and that man would gain access to them. There was at the same time a conviction that it was impossible to deduce general laws except by the careful observation of facts.

A reform of agriculture had commenced. New vegetables were culti-vated. New implements of husbandry were employed. New manures were applied to the soil. Temple had proved that many delicate fruits, the natives of more favoured climates, might be grown on English ground. Medicine, which in France was still in abject bondage, had in England become a progressive science. Attention was drawn for the first time to sanitary care. The great plague of 1665 induced men to consider drainage and ventilation. No kingdom of nature was left unexplored. To that period belongs the chemical discoveries of Boyle and the botanical researches of Sloane. One after another, phantoms which had haunted the world through ages of darkness fled before the light. Astrology and alchemy became jests. Soon there was scarcely a county Bench that did not smile contemptuously when an old woman was brought before them for riding on broomsticks or giving cattle the murrain.

Napoleon and Science

An extract from Vincent Cronin's book *The View from Planet Earth*.

The astronomer Laplace presented Napoleon with a copy of his *Celestial Mechanics* in five volumes. The Emperor leafed through it.

'You have written this huge book,' he said, 'without once mentioning the author of the universe?'

'Sire, I had no need of *that* hypothesis,' replied Laplace.[1]

Napoleon's world-pictures — for he held several in turn — are of some interest and typical of the France of his day. As a boy in Ajaccio he held the usual Catholic view of the heavens. At military academy in Burgundy he learned a lot of mathematics but less science than an English schoolboy learned, and nothing about astronomy — in a Catholic country now an awkward topic. As a subaltern he became keen on Rousseau. 'Go down to the sea shore,' he wrote at this time. 'Look at the morning star as it sets majestically on the breast of the infinite. Melancholy will overcome you. None can resist the melancholy in Nature.'

Napoleon dropped this emotional response to landscape and sky under the influence of the cult of Reason. As a leading general of the Republican armies he counted among his enemies the Catholic Church, both as a

199

temporal power and in the role it had adopted to Galileo. He would have described himself then as an agnostic. Yet, sailing to Egypt, he had lain on deck, asking his scientists whether the planets were inhabited, how old the Earth was, and whether it would perish by fire or by flood. Many, like his friend Gaspard Monge, the first man to liquefy a gas, were atheists. Then Napoleon would point to the stars. 'You may say what you like, but who made all those?'

When he became First Consul and started reconstructing, Napoleon saw that belief in the Christian God was a fact of life in France and, though he could not share the popular belief, he did restore the Catholic Church. But he had no time for an unscientific interpretation of nature, whether clerical or lay. When Bernardin de Saint-Pierre complained that scientists scoffed at his sentimental reading of nature, Napoleon asked sharply: 'Do you understand differential calculus, Monsieur Bernardin? Well then, go and study it.'

Yet when Laplace produced his nebular hypothesis, Napoleon found its atheism difficult to swallow.[2] It was more than esteem for his devoutly Catholic mother, more than his experience of how men can behave when they throw God overboard. Napoleon had pondered the perennial human perplexities: how the spirit functions through the flesh which at the same time seems to burden its activity. Yet any incipient belief in God did not deter Napoleon, in 1814, from trying to commit suicide by swallowing poison.

A prisoner on the rainswept rock of St. Helena, in December, 1817, Napoleon said to his companion Gourgaud:

'The intelligence watching over the movement of the stars — only a property of matter, you know — do you really believe that it watches over men's actions too, and keeps account of them?'

'Sire, I believe in God, and would be very unhappy to be an atheist.'

'Bah! Look at Monge and Laplace. Vanity of Vanities!'

Napoleon enjoyed getting a rise out of Gourgaud, just as he did out of the atheist General Bertrand by praising the Catholic Church. The truth is, he was still divided. He felt sympathy with the scientists, but what then of courage and glory and France — all those values which became mere vapour if life had no meaning? And divided Napoleon died — sending for a Catholic priest, but still unsure whether man was just a spin-off from a nebula.

Notes

1. Laplace was briefly Napoleon's Minister of the Interior, but was furiously sacked for 'introducing infinitessimals into administration'.

2. The 'nebular hypothesis' of Laplace was essentially correct. He argued that the sun and planets had been formed from a huge cloud of gas and

dust. He did not know, of course, that this cloud was propelled by a shockwave of the violent explosion of a supernova star, and that all the atoms in our bodies were created in the nuclear furnace of its core.

A Panting Recitation

Only one scientist has ever tested safety in the mines by reciting Shakespeare. Lawrence Altman tells the story:

J.S. Haldane, father of the biologist J.B.S. Haldane, passed on to his son a talent for science and the tradition of self-experimentation. As a youth, Jack took deep-sea dives with his father, descended into mines, and climbed Pike's Peak in Colorado. He recalled one event in a pit in North Staffordshire:
  'After a while we got to a place where the roof of the mine was about eight feet high and a man could stand up. To demonstrate the effects of breathing firedamp, my father told me to stand up and recite Mark Anthony's speech from Shakespeare's *Julius Caesar*, beginning: 'Friends, Romans, countrymen.'
  I soon began to pant, and somewhere about 'the noble Brutus' my legs gave way and I collapsed on to the floor, where, of course, the air was all right. In this way I learnt that firedamp is lighter than air.' (The air near the roof was full of methane, or firedamp, which is a gas lighter than air, so the air on the floor was not dangerous.)

But Who Are You?

Robert Bunsen, the nineteenth-century inventor of the laboratory Bunsen burner, was a highly absent-minded scientist. Henry Roscoe tells this story of him:

Bunsen had a well-known difficulty in remembering names. One day a visitor called whom he knew quite well was either Strecker or Kekulé.
  During his conversation he tried without success to make up his mind which of these two gentlemen was his caller. First he thought it was Kekulé, then he convinced himself that he was talking to Strecker. At last, however, he decided that it was really Kekulé. So when his visitor rose to take leave, Bunsen, feeling confidence in this last conclusion, could not

refrain from remarking: 'Do you know that for a moment I took you for Strecker!'

'So I am!' replied his visitor in amazement.

The Conquest of Smallpox

Smallpox used to be one of the world's most dreaded diseases. It has now disappeared. Lawrence Altman explains why:

Smallpox, like many other viral infections, could not be cured, nor could it be effectively treated once it came on. But it could be prevented.

The weapon against it was a vaccine developed by Edward Jenner in 1796. His discovery became a model for all the immunizations used in medicine today: children no longer risk choking to death from diphtheria or suffering brain damage from a complication of measles, or being paralyzed from polio. No longer are people in danger of getting 'lockjaw' (tetanus) from stepping on a rusty nail; and many fewer babies are gasping for breath from the severe paroxysms of whooping cough.

Jenner, an English country physician, took his cue from milkmaids who noted that people who came down with cowpox often escaped later infection from smallpox. He spent years developing a vaccine from the sores of cowpox patients. He was encouraged by John Hunter, a surgeon to King George III, who, in a dramatic self-experiment, had purposely injected himself with pus from a patient with gonorrhea. Jenner innoculated eight year-old James Phipps with cowpox. Young Phipps got cowpox and recovered. Then, in the next and most crucial step, Jenner innoculated the boy about twenty times over the next two decades, but Phipps did not get the disease.

When Jenner published a report on his smallpox vaccine, doctors in other countries began using his technique. The grandfather of all vaccines worked because the cowpox and the smallpox viruses were so closely related that when cowpox virus was injected into a human, it tricked the body's immune system into producing antibodies that defended against future invasion, not only by the cowpox virus, but also by the smallpox virus. It was an unusual way to make a vaccine; most of the ones developed since have been based on the disease-causing organism itself, being either weakened during growth in eggs or cells in test tubes, or killed by chemicals, before being used to immunize.

'Microbes are a Menace'

The foremost discoverer of the causes of disease was Louis Pasteur. Paul de Kruif tells his story:

In 1831, microbe hunting had come to a standstill while other sciences were making great leaps. Marvellous microscopes were being devised, but no one had used them to examine the powers of sub-visible animals. There was no hint that these wretched beings were more efficient murderers than the guillotine or the cannons of Waterloo.

One day in that year, a nine-year-old boy ran frightened from a blacksmith's shop in a French mountain village. He had heard the hiss of a white hot iron on human flesh. The victim being treated had just been bitten by a mad wolf. The boy was Louis Pasteur.

Eight more victims of the mad wolf died of the throat-parched agonies of hydrophobia.* Their screams rang in the ears of this boy.

'What makes wolves mad, father?' he asked. 'Why do people die when they bite them?' His father, an old sergeant of Napoleon, had seen thousands of men die from bullets, but he had no notion of why people die from disease. 'Perhaps a devil got into the wolf,' he replied. That answer was as good as any from the wisest scientist or the most expensive doctor of the time. Indeed, the cause of all disease was mysterious.

No one knew anything of microbes. The Swede Carolus Linnaeus, who toiled at putting all living things in a vast card catalogue, threw up his hands at the very idea of studying them. 'They are too small, too confused, no one will ever know anything about them. We will simply put them in the class of Chaos.'

But by the time Pasteur had grown up to become a chemist, microbes were beginning to come into serious notice. He was made Dean of Sciences in Lille at the age of 26, and it was here that he stumbled on microbes. Creating a great wave of excitement that went on for 30 years, he showed the world how important microbes were.

Pasteur was told by the merchants of Lille that highbrow science was all right, but —.

*   Rabies.

'What we want to know, professor, is, does it pay? Raise our sugar yield from our beets and give us a bigger alcohol output, and we'll see that you and your laboratory are taken care of.'

Pasteur understood that science has to earn its bread and butter, and he started to make himself popular by giving thrilling lectures.

'Where will you find a young man whose curiosity will not immediately be awakened when you put into his hands a potato, and when with that potato he may produce sugar, and with that sugar alcohol?' he shouted enthusiastically one evening to an audience of prosperous manufacturers and their wives. Then one day M. Bigo, a distiller of alcohol from sugar beets, came to him in distress. 'We're having trouble with our fermentation,' he complained. 'We're losing thousands of francs every day. Will you come to our factory and help us?'

Pasteur hastened to the distillery in question and sniffed at the vats that were sick, that wouldn't make alcohol. He carried off some samples of the grayish slimy mess in bottles, together with some of the best beet pulp from the healthy foam vats where great amounts of alcohol were being stored. He knew nothing of *how* sugar ferments into alcohol — indeed, no chemist knew anything about it. Back in his laboratory, he decided first to examine the stuff from the healthy vats. He put a drop of it under his microscope, and found it was full of tiny, yellowish globules, and their sides were full of a swarm of curious dancing specks.

Some were in bunches, others in chains, and some had queer buds sprouting from their sides as if from infinitely tiny seeds.

'These yeasts are alive. It must be the yeasts that change beet sugar into alcohol!' he cried. But what can be the matter with the stuff in the sick vats?' He put a drop from the bottle that held the stuff from the sick vat beneath his microscope.

'But there are no yeasts in this one. There is nothing here but a mass of confused stuff. What does this mean?' At last a strange look about the juice forced its way to his attention. 'Here are little gray specks sticking to the walls of the bottle, and some more floating to the surface. But there aren't any specks in the healthy stuff where there are yeasts and alcohol. What can that mean?' he pondered. Then he fished down into the bottle and put a speck into a drop of pure water. He put this beneath his microscope.

His moment had come.

There was no yeast globes here, but something utterly strange, great tangled dancing masses of tiny, rod-like things, some of them alone, some of them drifting along like strings of boats, all of them shimmying with a weird, incessant vibration. He hardly dared to guess at their size. They were much smaller than the yeasts. They were only one 25 thousandth of an inch long!

That night he forgot about Bigo. Nothing existed for him but these strange dancing rods. In every one of the grayish specks he found millions of them.

He set up apparatus that made his laboratory look like an alchemist's den. He found that the rod-swarming juice from the sick vats always contained the acid of sour milk — and no alcohol. Suddenly a thought struck him. 'Those little rods in the juice of the sick vats are alive, and it is *they* that make the acid of sour milk, just as the yeasts must be the ferment of the alcohol!' He was now sure he had solved the 10,000 year-old mystery of fermentation, that he could *prove* that the little rods were alive, and that despite their miserable littleness they did giant's work, changing sugar into lactic acid.

'I shall have to invent some kind of clear soup for these rods that I think are alive, so that I can see what goes on,' he decided. 'They must have special food, and then I shall see if they multiply, if they have young, if a thousand of the small dancing beings appears where there was only one before.' He tried putting some of the vats into pure sugar water, but they refused to grow in it. The rods needed a richer soup.

After many failures he devised such a soup. Taking some dried yeast and boiling it in pure water, he added some sugar and a little carbonate of chalk to keep the soup from being acid. Then, on the point of a fine needle, he took some specks of juice from a sick fermentation. Sowing them in his new clear soup, and putting the bottle in an incubating oven, he waited anxiously for results.

The next day he held it up to the light. 'Something is changing here,' he noted. 'Rows of little bubbles are coming up from some of the gray specks I sowed yesterday. There are many new gray specks, all of them sprouting bubbles.'

At last he put a drop from his bottle beneath the microscope. Eureka! The field of the lens swarmed and vibrated with millions of the tiny rods. 'They multiply! They are alive!' he whispered.

Time and time again he did the same experiment, putting a tiny drop from a flask that swarmed with rods into a fresh clear flask of yeast soup that had none at all. Every time the rods appeared in billions, and each time they made new quantities of the acid of sour milk. Then Pasteur burst out to tell the world. He told his classes about his great discovery that such infinitely tiny beasts could make acid of sour milk from sugar. Here was the one important fact:

It is living things, sub-visible living things that are the real causes of fermentations.

Becoming director of scientific studies in the Normal School in Paris, he became convinced that other kinds of small beings did a thousand useful and perhaps dangerous things. He now set about his first attempts at showing that microbes are the real murderers of the human race. He dreamed that just as there is putrid meat, so there are putrid diseases. He tells how he suffered in this work with rotten meat, and of the bad smells that filled his laboratory. 'I have resolved to devote myself to these studies without thought of their danger, he said, adding the quotation from Lavoisier: 'Public usefulness ennobles the most disgusting work.'

After many dangerous experiments, during which Pasteur became ill and nearly died, he established the link between microbes and diseases. His heroism thrilled calm scientists. He made a popular speech at the Sorbonne that sent his audience home in worry. He showed them lantern slides of a dozen different kinds of germs. He darkened the hall and shot a single bright beam of light through the blackness.

'Observe the thousands of dancing specks of dust in the path of this ray!' he cried. 'The air of this is filled with these specks of dust, these thousands of little nothings that you should not despise, for sometimes they carry disease and death, the typhus, the cholera, the yellow fever and many other pestilences.'

'Lousy' School Books

The late Richard P. Feynman, professor of theoretical physics and winner of a Nobel Prize, was asked to judge school science books. His account of the experience is both hilarious and horrifying:

I was asked to serve on the State Curriculum Commission, which had to choose schoolbooks for the state of California. You see, the state had a law that books used in the public schools had to be chosen by the State Board of Education, so the board has a committee to advise them which books to take.

I agreed to join the committee.

Immediately, I began getting letters and telephone calls from publishers. They said things like: 'We're very glad you're on the committee, because we really wanted a scientific guy . . .' and: 'It's wonderful to have a scientist on the committee, because our books are scientifically oriented . . .' But they also said things like: 'We'd like to explain to you what our book is about . . .' and: 'We'll be very glad to help you in any way we can to judge our books . . .'

That seemed to me crazy. I'm an objective scientist, and it seemed to me the only thing the kids in school are going to get is the books, and *extra* explanation from the publishers was a distortion. So I always replied: 'I'm sure the books will speak for themselves.'

I learned how the commission normally rated new books. Its members would receive many copies of each book and they would give them to various teachers and administrators. Then they would get reports back from these people. Since I didn't know many teachers, I chose to read all the books myself.

A few days later a guy from the book depository called me and said: 'We're ready to send you the books, Mr Feynman. They weigh three hundred pounds.'

I was overwhelmed.

'It's all right, Mr Feynman. We'll get someone to help you read them.'

I couldn't figure that out. You either read them or you don't read them. I had a special bookshelf put in my basement study, and began reading all the books that were going to be discussed in the next meeting.

It was a pretty big job. My wife says it was like living over a volcano. It would be quiet for a while, then all of a sudden, 'BLLLLOOOOOOWWWWW!!!!'

The books were so lousy. They were false. They were hurried. Their authors would try to be rigorous, but they would use examples (like cars in the street for 'sets') which were almost okay, but in which there were always some subtleties. The definitions weren't accurate. Everything was a little bit ambiguous — the authors didn't know what was meant by 'rigour.' They were faking it. They were teaching something they didn't understand and which was useless for the child.

They would talk, for example, about different bases of numbers — five, six, and so on. That would be interesting for a kid who could understand base ten — something to entertain his mind. But what they had turned it

into was that every child had to learn another base! And then the usual horror would come: 'Translate these numbers, which are written in base seven, to base five.' Translating from one base to another is an utterly useless thing. There's no point to it.

Finally I come to a book that says: 'Mathematics is used in science in many ways. We will give you an example from astronomy, which is the science of stars. Red stars have a temperature of five thousand degrees, yellow stars have a temperature of five thousand degrees . . .' — so far, so good. It continues: 'Green stars have a temperature of seven thousand degrees, and violet stars have a temperature of . . . (some big number).' There are no green or violet stars, but the figures for the others are roughly correct. It's vaguely right — but already, trouble!

The book says: 'John and his father go out to look at the stars. John sees two blue stars and a red star. His father sees a green star, a violet star, and two yellow stars. What is the total temperature of the stars seen by John and his father?' — and I would explode in horror.

There were perpetual absurdities like that. There's no purpose whatever in adding the temperatures of two stars. It was all a game to get you to add, but they didn't understand what they were talking about. It was like reading sentences with a few typographical errors, and then suddenly a whole sentence is written backwards.

Then came my first meeting. The other members had given ratings to some of the books, and they asked me for mine. My ratings often differed from theirs, and they would ask: 'Why did you rate that book low?'

I would say the trouble with the book was this and this on page so-and-so — I had my notes.

I was asked what I thought of a certain book, part of a set of three published by the same company.

I said: 'The book depository didn't send me that book.'

The man from the book depository said: 'Excuse me; I can explain that. I didn't send it because that book hadn't been completed yet.' There's a rule that every book must be sent in by a certain time, and the publisher was a few days late. So it was sent to us with just the covers and all the pages blank. The company sent a note excusing themselves and hoping they could have their set of three books considered, even though the third one would be late.

But the blank book had a rating by some of the other members!

I believe the system works like this: when you give books all over the place to people they're busy; they're careless; they think: 'Well, a lot of people are reading this book, so it doesn't make any difference.' And they put in some kind of number. Then, when you receive your reports, you average the ratings. This process can miss the fact that there is absolutely nothing between the covers of a book!

This embarrassed them, and it gave me a little more confidence. It turned out that the other committee members had gone to sessions in which the book publishers would explain the books before they read them; I was the only one who read all the books and didn't get any information from the book publishers themselves.

This question of trying to figure out whether a book is good or bad by looking at it carefully or by taking the reports of a lot of people who looked at it carelessly is like a famous old problem: Nobody was allowed to see the Emperor of China, and the question was, what is the length of his nose? To find out, you go all over the country asking people what they think its length is, and you average it. And that would be very 'accurate' because you averaged so many people. But it's no way to find anything out. You don't improve your knowledge of the situation by averaging.

What finally made me resign was that the following year we were going to discuss science books. I thought maybe they would be different, so I looked at a few of them.

The same thing happened: something would look good at first and then turn out to be horrifying. For example, there was a book that started out with four pictures: first there was a wind-up toy; then there was a car; then there was a boy riding a bicycle; then there was something else. And underneath each picture it said: 'What makes it go?'

I thought, 'I know what it is: they're going to talk about mechanics, how the springs work inside the toy; about chemistry, how the car engine works; and biology, about how the muscles work.'

It was the kind of thing my father would have talked about: 'What makes it go? Everything goes because the sun is shining.' And then we would have fun discussing it:

'No, the toy goes because the spring is wound up,' I would say.

'How did the spring get wound up?' he would ask.

'I wound it up.'

'And how did you get it moving?'

'From eating.'

'And food grows only because the Sun is shining. So it's because the sun is shining that all these things are moving.' That would get the concept across that motion is simply the transformation of the sun's power.

I turned the page. The answer was, for the wind-up toy, 'Energy makes it go.' For everything. 'Energy makes it go.'

Now that doesn't mean anything. Suppose it's 'Wakalixes.' That's general principle: 'Wakalixes makes it go.' There's no knowledge coming in. The child doesn't learn anything; it's just a word!

It's also not even true that 'energy makes it go,' because if it stops, you could say, 'energy makes it stop' just as well. What they're talking about is concentrated energy being transformed into more dilute forms, which is a very subtle aspect of energy. Energy is neither increased nor decreased in these examples; it's just changed from one form to another.

But that's the way all the books were. They said things that were useless, mixed-up, ambiguous, confusing, and partially incorrect. How anybody can learn science from these books, I don't know, because it's not science.

So when I saw all these horrifying books, I saw my volcano process starting again. Since I was exhausted from reading all the books, and discouraged from it's all being a wasted effort, I couldn't face another year of that, and had to resign.

Sometime later I heard that the energy-makes-it-go book was going to be recommended by the curriculum commission, so I made one last effort. At each meeting the public was allowed to make comments, so I got up and said why I thought the book was bad.

The man who replaced me on the commission said: 'That book was approved by sixty-five engineers at the Such-and-such Aircraft Company!'

I didn't doubt that the company had some pretty good engineers, but to take sixty-five engineers is to take a wide range of ability — and necessarily to include some pretty poor guys! It was once again the problem of averaging the length of the Emperor's nose, or the ratings on a book with nothing between the covers. It would have been far better to have the company decide who their better engineers were, and to have them look at the book. I couldn't claim that I was smarter than sixty-five other guys — but the average of sixty-five other guys, certainly!

I couldn't get through to him, and the book was approved.

Whenever our commission had a meeting, there were book publishers entertaining members by taking them to lunch and talking to them about their books. I never went.

I once received a package of dried fruit and whatnot with a message that read: 'From our family to yours, Happy Thanksgiving — The Pamilios.'

It was from a family I have never heard of, obviously someone who had the name and address wrong, so I thought I'd better straighten it out. I telephoned the people who sent the stuff.

'Hello, my name is Feynman. I received a package —'

'Oh, hello, Mr Feynman, this is Pete Pamilio' and he says it in such a friendly way that I think I'm supposed to know who he is.

I said, 'I'm sorry, Mr. Pamilio, but I don't quite remember who you are.'

It turned out he was a representative of one of the school book publishers.

'I see. But this could be misunderstood.'

'It's only family to family.'

'Yes, but I'm judging a book that you're publishing, and maybe someone might misinterpret your kindness!' I knew what was happening, but I made it sound like I was a complete idiot.

Another thing like this happened when one of the publishers sent me a leather briefcase with my name nicely written in gold on it. I gave them the same stuff: 'I can't accept it; I'm judging some of the books you're publishing. I don't think you understand that!'

But I really missed one opportunity. If I had only thought fast enough, I could have had a very good time on that commission. I was at a hotel in San Francisco before my very first meeting the next morning, and I decided to wander in the town and eat something. I came out of the lift, and on a bench in the lobby were two guys who jumped up and said: 'Good evening Mr. Feynman. Where are you going? Is there something we can show you in San Francisco?' They were from a publishing company, and I didn't want to have anything to do with them.

'I'm going out to eat.'

'We can take you out to dinner.'

'No, I want to be alone.'

'Well whatever you want, we can help you.'

I couldn't resist. I said, 'Well, I'm going out to get myself in trouble.'

'I think we can help you in that, too.'

'No, I think I'll take care of that myself.'

Then I thought, 'What an error! I should have gone with them and kept a diary, so the people of California could find out how far the publishers will go!'

Sharks

Vane Ivanovic, diplomat and shipowner, is a renowned expert on these terrible animals:

I have found sharks an impossible subject for drawing room conversation. When I say: 'I am not afraid of sharks and I dive among them,' I am taken for a madman or a liar. If I say: 'I am very much afraid of sharks, but still dive among them,' I am thought to boast or to exhibit an even more disturbed mind. Yet both statements are true. I was once very much afraid of sharks. Now I am not at all afraid of most sharks, only of some.

I have myself encountered 344 sharks. Not a single one of them ever made an aggressive move towards me — not even the twelve I wounded, of which I successfully landed seven. We know far too little about sharks to make any definite statements on shark behaviour. To call them unpredictable is a roundabout confession of our ignorance. It would be wiser to admit it, rather than to ascribe qualities to sharks that they may not possess.

Sharks are not modern animals like tigers and lions, to name but two of the land beasts known to attack human beings and eat their flesh. Sharks belong much more to the age of dinosaurs in their balance between instinct and deduction from experience. They have had the luck to remain utterly superior to any possible enemy, and have thus maintained themselves in their environment under the sea. They have survived in their state of superb brutal strength, capable of fast movement and of only the most primitive cunning. Relatively few sharks have come against the only modern creature capable of destroying them — man. I say 'relatively few,' considering the millions of people who have been in or on the sea, and the millions of sharks known to exist. Fortunately for sailors, bathers and shipwrecked seamen, the vast majority of sharks rarely come near the surface of the sea.

Sharks near the surface, when we have been able to observe them, do not attack living fish capable of taking care of themselves. They will attack dead or dying fish. Sharks are attracted by anything that causes unusual vibrations at or near the sea's surface — vibrations that seem unusual for the undersea world. A tea-kettle bobbing about on the surface or an old boot gradually sinking will convey messages to the vibration-sensing mechanism on the shark's skin — messages as promising of a good meal as the vibrations given off, say, by dead or dying sheep thrown into the sea.

The shark will sense from great distances something moving unnaturally in the water and will rush to the spot. Sharks will often strike without circling around to inspect the object to be attacked. They will have decided to attack long before they have smelt or seen blood. Every trolling fishermen knows that he can attract a shark by dangling a simple piece of cloth at the end of his line. As long as this cloth is weighted so that it will bob up and down at the surface and move at a speed at which wounded mackerel might travel, the shark will attack it without pausing to smell flesh or see blood.

I call most shark attacks on human beings *provoked* attacks. Human legs, dangling like two luscious bananas from a lifebelt, are highly provocative objects for a shark. The human body, swimming freely with head out of the water, or a person standing or splashing about in shallow water, are an irresistible provocation to shark-attack.

A spear-fisherman wearing a mask and rubber fins and swimming along smoothly, even on the surface, will give a shark the impression of a large fish very much alive and able to take care of itself. Very often I have dived towards a shark, giving every appearance of being about to attack it. Without exception, they have disappeared at once. The smallest man wearing rubber fins is about six feet long in the water. That makes quite a sizable fish, as fish go. Of course a spear-fisherman carrying with him dead or dying fish may be attacked. Sharks are pretty stupid. When they appear from out of the blue in response to vibrations of dying fish, they cannot tell exactly where the vibrations are coming from. I have been lucky in that the only time a shark rushed at me was to attack a fish still quivering at the end of my long spear. It got the fish, but not me or my rubber fins.

I think I have read almost all the books and articles written about sharks in the English language during the last quarter century. With a few honourable exceptions they are the propagation of hearsay and the fantasies of braggarts. Even a serious American scientific study allows its rational procedures to collapse when it reached its conclusions.* After carefully sifting their evidence, the authors conclude that since the beginning of the world, only just under 400 cases of death due to shark attack could be considered as proven. Of course, evidence of shark attacks in ages past had been rightly dismissed as unreliable. But then, having established these fairly recent cases, the authors conclude that the coasts of Australia are the most dangerous, since most proven shark attacks had taken place there. They say that sharks evidently attack most often in the hot season in January and February. But they fail to point out that Australia is the only place in the world where, in a tropical or subtropical climate, literally millions of people bathe in these months. There are therefore many more potential victims and also many witnesses. Of course Australian sharks do not attack swimmers in August and September, because few people in Sidney or Melbourne are crazy enough to bathe in the winter.

When flying over warm seas, I always carry a mask and a pair of rubber fins under my seat. If the aircraft came down, my best chance of survival in tropical waters would be to remain swimming in a horizontal position, pretending to be a big, live, if rather lethargic fish.

I have tried to persuade the Captains and crews on my ships to carry in the ship's lifeboats masks and rubber fins as additional life-saving equipment. If I were doomed to spend days in a lifeboat, with a relentless tropical sun beating down on me and dehydrating me, I would slip into the sea many times a day and spend many daylight hours swimming, while holding on to the lifeboat to avoid losing contact. For this purpose, and to avoid shark attack, I would prefer to wear a mask and a pair of rubber fins.

On the subject of sharks, as on several other subjects, my reasoning and experience are regarded as rather eccentric.

---

* Material collected in *Shark Attack*, by V.M. Coppleson (1958).

The Baron and the Thief

Only one scientist has ever been expelled from Fellowship of the Royal Society of London for infamous conduct. This was Rudolph Raspe, adventurer and thief, who was famous also for writing the fabulous and mendacious adventures of Baron Munchausen. John Carswell tells the tragic fate of the real-life Baron and military veteran at the hands of the unscrupulous Raspe.

From the beginning, as a student at Göttingen University, Raspe was short of money. He began to accumulate what he later called 'debts from zeal for learning.' He developed wide and well-informed antiquarian and artistic tastes; but he was more scientist than aesthete, and better fitted to classify antiques than to appreciate them.

In 1763 he published an ambitious work on volcanic geology which soon gained him international fame. It earned the praise of Sir Charles Lyell eighty years later, who called it a work of 'luminous exposition.' He was taken up by the pundits in the university, and he responded with all the devotion of an ambitious man. Appointed secretary of the State Library of Hanover, he soon found a distinguished patron in Count Walmoden, owner of the largest artistic collection in the Electorate.

The energetic librarian was delighted with a commission to catalogue the galleries and cabinets. His life began to expand. He attended masquerades and balls with the 'beautiful and agreeable' wife of his cousin, von Einem, who bided his revenge. Borne on a rising tide, he allowed himself the luxury of strident polemic. He attacked his critics both publicly and privately with a caustic wit that earned him little affection.

Supreme good fortune was now at hand. In 1767, the 'trusty and well-beloved Rudolph Erich Raspe' was appointed Professor of Antiquity and Keeper of the Collections to Frederick II, Landgrave of Hesse-Cassel.

Raspe's brain hummed with projects. He catalogued and rearranged the Landgrave's collections, discovering some 600 hitherto unlisted items. He now asked for an advance of capital, to clear himself of creditors who grew daily more vociferous. His European reputation was costing him a tenth of his salary in postage alone, and he owed about three years' income to moneylenders.

His energy did not slacken. He had long sought election as a Fellow of the Royal Society of London, and to this end he cultivated any distinguished Englishman who came within range. He was elected in 1769, and rejoiced in the 'entirely unsolicited honour.'

High posts and learned works might bring honour, but they brought few rixthalers. He applied to the Prime Minister of Hesse but received only soothing answers. Only marriage was left; and Raspe chose for his bride the daughter of a wealthy Berlin doctor. When the ceremony was over, he was the richer by 2,000 rixthalers. His credit and spirits temporarily revived.

But the assistance came too late. The moneylenders were threatening him with a bankruptcy that would ruin his European reputation. He plunged into fresh projects. He began in 1772 to edit a chatty periodical called the *Kassel Spectator*; it reached its 24th number and collapsed. He borrowed 2,000 rixthalers from his father-in-law, but to no one did he admit the whole truth — that since 1770 he had embezzled the medals in his charge.

In 1774 a last chance seemed to offer an escape. He was invited to be Hessian resident in Venice. The salary was not much larger, but the prospects for promotion were good. He accepted eagerly. He secured a substantial advance for travelling expenses, which he threw to his creditors. While they swarmed with fury at the news of his impending departure, he set out for Berlin to borrow a very large sum from his father-in-law. For there was one snag in the Venetian project: what would happen when he handed over the keys of the collection to his successor?

The authorities now demanded his presence at an inventory. He returned to Hesse. His fraud was exposed from the very catalogue which he had himself compiled. He confessed in the most abject terms, vainly imploring the Landgrave's forgiveness. Raspe was posted as a fugitive from justice throughout North Germany.

Fleeing to Holland, he wrote an apologia full of self-pity and ingenious special pleading. It only served to revive the flagging animosity of the Landgrave. His pay, he explained, had been long in arrears. He had chosen medals for the pawnbroker with the utmost discrimination, taking only those which lacked aesthetic or antiquarian interest. Such remarks explain why Raspe's name was remembered in Hesse to the end of the century as shameful. His enemies hastened to destroy what was left of his reputation. His cousin von Einem, whose wife Raspe had seduced, even suggested that he had replaced the Landgrave's jewellery with paste.

Absconding to England, Raspe began to make the rounds of his colleagues in the Royal Society. But the truth soon overtook him. Its president, Sir John Pringle, was anxiously seeking confirmation of rumours about Raspe's past. Two days later, Raspe was expelled from the Royal Society by a large majority.

His hopes were now blighted, but for the rest of his career he showed a courage which he had lacked in the days of his better fortune. He succeeded in installing himself as master of an office in the Cornish tin belt. Here he experimented, prospected and wrote reports. He learned the jargon of the Cornish miners, and showed them that he knew more about practical mining than they did.

Yet he did not neglect letters. He made the acquaintance of Smith, an Oxford printer, who liked his stories of a half-remembered figure of his youth, Baron Karl von Munchausen, the veteran cuirassier of Bodenwerder.*

Raspe must have enjoyed writing about Munchausen. It was the work, one can imagine, of a few summer evenings, and is charged with that light-hearted malevolence with which the author revenged himself on the world. There is more than a merely accidental contrast between Munchausen, the fabulously successful man of action, and Raspe, the seedy failure; between the Baron who strikes his hearers into dumb amazement, and the learned courtier who has never quite been able to make himself heard at all. Raspe's conscious intention was satire, aimed at enemies who had, he persuaded himself, ruined his career. But more obscurely the Baron was the creation of his own inflated ego, and the real villain of his tragedy.

Baron Munchausen took his place on the world stage in 1785 in a 42-page pamphlet published by Smith called *Baron Munchausen's Narrative of his Marvellous Travels and Campaigns in Russia*, and in the following year there was a new edition called *Gulliver Revived: The Singular Travels, Campaigns, Voyages and Sporting Adventures of Baron Munnikhousen, commonly pronounced Munchausen; as he related them over a Bottle of Wine when surrounded by his friends*. It was soon translated into German with elaborations.

Towards the end of his life, Baron Munchausen's peaceful and harmless existence was thus violently interrupted by his sudden elevation to the status of a legend. He grew embittered and morose. Too self-conscious to continue it, he abandoned the straight-faced humour with which he had once held his dinner-table in helpless laughter. He remarried at the age of 74 in an attempt to regain some consolation in a Bodenwerder from which his faithful huntsman, Rosemeyer, in vain tried to exclude the sightseers. But his new wife was a hussy, with her eye on the old squire's estate. The once genial raconteur died in 1797, an unhappy man.

It is perhaps Raspe's gravest crime to have ruined the happiness of the defenceless old soldier. To combine his gleanings with some of the Baron's own sporting adventures, heard perhaps twenty years before at a

---

* Baron von Munchausen was no fictitious character, as many people imagine. He was a passionate hunter and soldier, who served for many years with the Russian armies against the Turks. He retired to his estate in Bodenwerder in 1760, where telling amusing after-dinner stories became his favourite occupation.

dinner-party at Bodenwerder, and to embellish both with his own fantasies, must have seemed an easy way of earning a few guineas to a man accustomed to laborious and unprofitable work. In the process he left on his hero the imprint of his own disreputable and egocentric genius.

Heaven is Hotter than Hell*

The light of the Moon shall be as the light of the Sun, and the light of the Sun shall be sevenfold, as the light of seven days.
—Isaiah 30,26.

Thus heaven receives from the Moon as much radiation as we do from the Sun, and in addition seven times seven (49 times) as much as the Earth does from the Sun, or some 50 times in all. With this data, we can compute the temperature of Heaven. The radiation reaching Heaven will heat it to the point where the heat lost by this same radiation is just equal to the heat generated by the radiation it receives. In short, Heaven loses fifty times as much heat as does the Earth.

Using the 19th century Stefan-Boltzman law which predicts that the radiation emitted by an object is proportional to the fourth power of its temperature, we can compute that the absolute temperature of the Earth is 80 degrees F (27°C).

The exact temperature of Hell cannot be computed, but it must be less than 230 degrees F (110°C.) the temperature at which brimstone or sulphur changes into a gas. According to Revelations 21,8: *The fearful and unbelieving . . . shall have their part in the lake which burneth with fire and brimstone.* A lake of molten brimstone means that its temperature must be below the boiling point, which is 230°F. (Above this point it would be a vapour, not a lake.)

We have, then, the temperature of Heaven at 274 degrees F. The temperature of Hell is less than 230 degrees F. Therefore Heaven is hotter than Hell.

---

* From an unsigned article in *Appled Optics*, Vol. 11, 1972.

The Explosions Within Us

What causes spontaneous cancers and cell mutations which can lead to deformed births? In this fascinating chemical detective story, Isaac Asimov was the first to draw attention to the sinister role of naturally occurring carbon–14.

Is it possible that radioactive atoms may find their way into living tissues and even into our own bodies?

It is not only possible; it is certain.

To determine the nature of the radioactivity in the human body, we must know the composition of living tissue. I therefore begin with a list of the elements that occur in it, and my best estimates as to the quantity of each present. You will find this in Table I.

You might wonder about the elements at the bottom of the table, the ones that occur only to the extent of a few atoms per billion. They are usually called the 'trace elements' because they are present only in traces. Does the body really need them? It certainly does. With the exception of fluorine, they are absolutely essential to human life and even fluorine is necessary for healthy teeth.

Does it seem strange to you that the body can do with so little and yet not be able to get along with none at all? From five atoms per billion to zero atoms per billion seems such a small step.

Well, it's all in the way you look at it. Suppose we count the atoms involved.

Start with a person weighing 150 pounds. He or she is made up mostly, but not entirely, of microscopic cells, the individual chemical factories of the body. The 'not entirely' part comes about as follows: in the blood and in the spaces between the cells there is a total of some 30 pounds of fluid (mostly water) which forms no part of any cell and is called 'extracellular fluid.' In the bones and teeth there are some 15 pounds of mineral matter which is also extracellular. This leaves 105 pounds of cells.

The average liver cell weighs about one fourteenth of a billionth of an ounce. Let's assume that this is about average for the weight of a cell. In that case, there are thus some twenty-five trillion cells in the body.*

---

* A trillion here means a million million, and a billion a thousand million. This is now the general style in popular scientific writing.

TABLE 1

Element	Number of atoms present for every billion atoms in the body.
Hydrogen	630,000,000
Oxygen	255,000,000
Carbon	94,500,000
Nitrogen	13,500,000
Calcium	3,100,000
Phosphorus	2,200,000
Potassium	570,000
Sulphur	490,000
Sodium	410,000
Chlorine	260,000
Magnesium	130,000
Iron	38,000
Zinc	1,500
Maganese	170
Copper	170
Fluorine	125
Iodine	20
Molybdenum	10
Cobalt	5

The material outside the cells does not contain the same elements in the same proportion as the material inside the cells. For instance, the extracellular fluid is richer in sodium and poorer in potassium than the material inside the cells. The mineral matter in the bones is richer in calcium and phosphorus, and poorer in carbon and nitrogen, than the material inside the cells.

Furthermore, the cells of various tissues differ among themselves. Liver cells, for example, have at least two or three times as high a concentration of copper and cobalt as most other types of cells; red blood cells are particularly rich in iron, and so on.

Nevertheless, to begin with, I am going to suppose that the material of the body is divided up evenly among the cells and the extracellular material. Well, then, each cell contains about ninety trillion atoms. Using Table I, it is easy to calculate how much of each trace element is present in each cell.* The figures are given in Table II.

---

* I leave out fluorine in Table II since it is not essential to life, and it only occurs in the mineral matter of the bones and teeth and hardly at all in the cells themselves. I.A.

TABLE 2

Element	Number of atoms present in each cell
Zinc	135,000,000
Manganese	15,300,000
Copper	15,300,000
Iodine	1,800,000
Molybdenum	900,000
Cobalt	450,000

So even in the case of the least of these trace elements, cobalt, each individual cell, each little factory of the body, has nearly half a million atoms at its disposal.

Now the difference between five per billion and zero per billion may not seem much; but certainly there is a vast difference between a cell having a few hundred thousand atoms and its having none at all.

Looking over the list of elements in the human body in Table I, we see at once that we can forget about any of the long-lived varieties of unstable atoms. All, that is, but one! That one is potassium—40.

The body contains 570,000 atoms of potassium in roughly every billion of atoms. One out of every 9,000 atoms of potassium is potassium—40, the radioactive variety. This means that out of every billion atoms in the body, 63 are potassium—40.

This is no small amount. There is more than three times as much potassium—40 in the body as there is iodine. If potassium is considered to be spread evenly through the body, there would be, on the average, about five and a half million atoms of potassium—40 per cell. Actually, it is worse than that. Ninety-eight per cent of the body's potassium is within the cells and only two per cent is in the extracellular material. That raises the number to an even eight million atoms of potassium-40 per cell.

Fortunately, all those atoms of potassium-40 aren't breaking down simultaneously. At any time, only a comparatively small number are breaking down since potassium-40 is a long-lived atom with a half-life of over a billion years. In fact only one atom out of every 53,000 trillion atoms of potassium-40 breaks down each second.

But don't heave a sigh of relief too soon. In the body as a whole there is so much potassium-40 that even at this incredibly low rate of breakdown, 38,000 atoms of potassium-40 are exploding each second. In nine tenths of the breakdowns, a beta particle is emitted. This means that every second, we are subjected to the effects of 35,000 beta particles criss-crossing within us. Things may seem a little better, though, if we consider the explosions in a single cell rather than in the whole body. Each particular cell undergoes one of these explosions, on the average, only once every 200 years.

Or, to put it still more comfortingly, if you live for 70 years, then the odds are two to one that any particular cell of your body will never know that it is to have a potassium-40 atom explode within it.

Well, how do these explosions affect us? Obviously, they don't kill us outright. We're not even aware of them.

Yet they have the capacity for damage. Enough radioactivity can kill and has killed, but 35,000 explosions per second are not nearly enough to do that.

What about milder effects, though? A beta particle, as it darts out of a breaking down potassium-40 atom, usually hits a water molecule (by far the most common molecule in the body) and knocks off a piece of it. What is left of the water molecule is called a 'free radical.' Free radicals are reactive substances that will tear into any molecule they come across.

There is always a chance, then, that the unfortunate molecule that finds itself in the path of a free radical may be one of the nucleo-protein molecules we call 'genes.' There are several thousand genes in each cell, each controlling some particular facet of the cell's chemistry. If one of those genes is damaged or altered as a result of a collision with a free radical, then the cell's chemistry is also altered to some extent. The same thing happens, of course, if the beta particle should happen to hit the gene directly.

If the cells whose chemistry is altered happen to be germ cells (that is, cells which eventually give rise to ova or spermatozoa — as the case may be), it is quite possible that the offspring of the organism exposed will end up with a chemical organization different from that of its parents. The change may be so small and unimportant as to be completely unnoticeable, or sufficiently great as to cause physical deformity or early death. In either case the change is a 'mutation.'

If a gene is changed in a cell other than a germ cell, the cell may be altered with no permanent hardship to the body as a whole, or (just possibly) it may be converted to a cancerous cell with very drastic results.

This is not just speculation. Animals exposed to high-energy radioactive particles, or to radiation energetic enough to be capable of manhandling genes, may be damaged to the extent of developing radiation sickness and dying. At lower doses, the animals will show an increased incidence of cancer and of mutations.

Nor are human beings immune. Radioactive radiations and X-rays have caused cancer in human beings and killed them, too. Some skin cancers have been attributed to over-exposure to the Sun's ultra-violet radiation.

But when all known causes of cancer and mutations are eliminated there always remain a certain number that seem 'spontaneous'; that arise from no known cause.

Well, then, the thought arises, or should arise — can these 'spontaneous' cases be due to the potassium–40 beta particles careering around within us?

The answer of this question seems conclusive. The effect is insufficient! The radiation to which a human being is subjected from potassium–40 atoms within him is about the same as the radiation which he gets from cosmic ray bombardment from outer space. Both put together cannot possibly account for more than a tiny fraction of the 'spontaneous' cancers and mutations.

But are we through? Have we exhausted the possibilities of explosions within us?

The answer is 'No' to both questions. The bombardment of the atmosphere by cosmic rays results in the continuous production of carbon–14. Carbon–14 is radioactive and has a half-life of only 5,570 years, but cosmic rays maintain its presence at an even level, at a rate that just balances its rate of breakdown.

This 'even level' is certainly not very high. Carbon atoms exist in the atmosphere in the form of gaseous carbon dioxide. Only 0.04 per cent of the atmosphere is carbon dioxide. Only one of the three atoms in the carbon dioxide molecule is carbon. And only one carbon atom out of every 800 billion is carbon–14. Certainly, the total amount of carbon–14 present in the atmosphere doesn't seem to be overwhelming. Let's see!

The weight of the atmosphere is 14.7 pounds on every square inch of the Earth's surface. Take into account all the square inches there are on the Earth's surface (a little over 800,000 trillion if you're curious) and the total weight of the atmosphere turns out to be 5,900 trillion tons.

From this we can see that the total weight of carbon dioxide in the atmosphere is 2,360,000 billion tons; and the total weight of the carbon atoms themselves is 750 billion tons; and the total weight of the carbon–14 is 0.9 tons, or 1,800 pounds. Indeed, not an overwhelming amount, but then not an insignificant one either.

Eighteen hundred pounds of carbon-14 contain 35,000 trillion trillion atoms. If these atoms are spread evenly through the atmosphere, each cubic inch of air (at room temperature and sea-level pressure) would contain 120,000 atoms of carbon–14. In the air contained in a moderate-sized living-room there would be over three hundred billion atoms of carbon–14.

Put it another way. Every time you breathe, 30 cubic inches of air moves in and out of your lungs. That means that each time you breathe, you pump three and a half million atoms of carbon–14 into your lungs. In an average lifetime, you will have breathed over two thousand trillion atoms of carbon–14.

Not insignificant at all!

But do any of these atoms become part of living tissue? They certainly do; that is the crux of the whole thing.

Living matter is approximately 10 per cent carbon by weight. Every bit of that carbon, whether the organism is large or microscopic, plant or animal, of the sea or of the land, came originally from the carbon dioxide of the air. Green plants incorporate carbon dioxide of the air into large carbon-containing molecules. Animals eat the plants (or other animals that have already eaten the plants) and use the carbon atoms as hand-me-downs for their own purposes.

Since living tissues makes practically no distinction between carbon–14, which is radioactive, and carbon–12 and carbon–13, which are both stable, the proportion of carbon–14 in living matter is the same as it is in air.

As long as you are alive, the proportion of carbon–14 in your body remains constant. You remain in constant balance with the atmosphere, where the production of carbon–14 by cosmic rays just balances its rate of breakdown. The result is that carbon isolated from any recently living object contains enough carbon–14 to liberate 450 beta particles a minute for every ounce of carbon present.

And so what about the effects of the carbon–14 in our body?

The body of the 150-pound person I mentioned earlier contains about 300 trillion trillion carbon atoms. Of these, a tiny fraction, some 350 trillion, are carbon–14 atoms. If we omit the mineral matter of the body, and assume the carbon–14 atoms are otherwise spread out evenly, then each cell contains just about 11 atoms of carbon–14.

This is quite a small figure when you compare it with even the rarest trace element in the body. There are 40,000 times as many cobalt atoms in a cell, at the very least, as carbon–14 atoms. The comparison with potassium–40 is even more extreme. There are over 700,000 times as many potassium–40 atoms in a cell as carbon–14.

If it has been decided, then, that potassium–40 is quite harmless to the body, it would certainly seem, thousands of times more emphatically, that carbon–14 ought to be harmless.

But wait! Carbon–14 has a much shorter half-life than potassium–40. An equivalently larger proportion of its atoms ought to be breaking down per second, and after all, it is the number of beta particles being produced, and not the number of atoms, that counts.

Well, knowing the number of carbon–14 atoms present and the half-life of carbon–14, we can calculate that each second there are some twelve hundred beta particles produced within the body by carbon–14 break-downs.

The proportion of potassium–40 is still greater, but no longer so one-sidedly. Potassium–40 produces nearly 30 times as many beta particles as does carbon–14.

But another point must be considered in addition to mere numbers. The energy of the beta particles produced by potassium–40 is some ten times as great as those produced by carbon–14. The potassium–40 atom, as it breaks down, can therefore do ten times the damage that carbon–14 atoms can do as they explode. That swings the pendulum in the other direction again, for it would now appear that potassium–40 in the body does, on the whole, 300 times the damage carbon–14 does.

Certainly, it seems that no matter what we do, or how we slice it, carbon–14 remains badly out of the running.

Ah, but carbon–14 has a trump up its sleeve!

The key molecules of the cells are its genes. It is change in the genes which can bring about mutations and cancer. These genes contain no potassium atoms in their molecule. Any potassium–40 atoms present in the cell are located elsewhere than in the gene. Beta particles that shoot out of an exploding potassium–40 atom must strike a gene molecule just right, or give rise to a free radical which will strike it just right. It is as though you were standing in a globular shooting gallery, blindfolded, and aimed at random in the hope of hitting a few tiny targets placed here and there on the walls, floor and ceiling.

All in all, the chances of a beta particle from a potassium–40 atom damaging a gene, either directly or indirectly, is very low; perhaps only one out of many millions.

But now let's consider carbon–14. The gene is not merely being shot at by beta particles from exploding carbon–14 atoms. It *contains* carbon–14 atoms.

The genes make up about 1 per cent of the average cell. This means there are some twenty-four trillion trillion atoms in the genes of which half, or 12 trillion trillion, are carbon atoms. Of these, some fourteen trillion are carbon–14 atoms.

It comes to this, then: There is, on the average, one carbon–14 atom in the genes of every two cells.

The number of these that break down can be calculated. It turns out that each second, in your body as a whole, 50 carbon–14 atoms, *located in the genes*, are exploding and sending out a beta particle. These may be weaker and fewer than the beta particles sent out by exploding potassium–40 molecules, but every one of these fifty scores a hit!

Even if you suppose that the beta particle from an exploding carbon–14 atom within a gene might plow through the remainder of the gene without hitting any of its atoms squarely enough to do damage, the fact still remains that the exploded carbon–14 atom has been converted to a stable nitrogen–14 atom. By this change of carbon to nitrogen, the gene is chemically altered (only slighly perhaps, but altered nevertheless). Furthermore, the carbon-14 atom, having shot out a beta particle, recoils just as a rifle would when it shoots out a bullet. This recoil may break it away from its surrounding atoms in the molecule, and this introduces another change, and an even more important one.

For potassium–40 to damage a gene as much as the gene's own carbon–14 will do, the chances of a beta particle from a potassium–40 atom (or its free radical product) striking a gene hard enough to do damage must be at least as good as one in 8,000. The chances just aren't that high or anything like it, so I end by concluding that carbon–14 is much more likely to be responsible for 'spontaneous' cancer and mutations than potassium–40 is.

And if that is so, there is precious little that can be done about it unless someone turns off the cosmic rays, or unless we build underground cities.

But the situation isn't as serious as you might think. Fifty explosions per second within your genes may sound as though you couldn't last very long without developing cancer or having deformed children; but remember, a change in a gene may mean *any kind* of change. In the vast majority of cases, any serious change simply results in that particular gene in the particular cell refusing to work at all. Probably only a vanishingly small percentage of the changes result in cancer. (And if we only knew what the details of that change was!)

Then again, some cells are more important than others. Only the cells that give rise to ova and spermatozoa can be responsible for mutated children, and they form only a small percentage of the total number of cells in the body.

Fifty explosions per second over the entire body doesn't mean much to the individual cell (any more than the fact that twenty stars in our Galaxy blow up each year should cause us to worry much about our Sun). If you were to consider a single particular cell in the body, chosen at random, then an average of 18,000 years must pass before a single carbon–14 explodes in its genes.

This is the same as saying that if you live to be 70, the chances that a particular cell in your body will ever have experienced even a single carbon–14 breakdown in its genes is only one in 260.

So sleep in comfort!

Note

This article was first written in 1956. Some two years earlier I had written a short article on the same subject which appeared in the February, 1955, issue (pages 84–85) of the *Journal of Chemical Education*. That, I believe, was the first mention in print of the relationship of carbon–14 to genetics.

In 1958, when atmospheric testing of nuclear bombs still went on wholesale, Linus Pauling (my favourite chemist) published a paper in *Science* (November 14, 1958) which pointed out the way in which testing would increase the carbon–14 content of the atmosphere and therefore the incidence of undesirable mutations.

I have a letter from Professor Pauling, dated 11 February 1959, which refers in most kindly fashion to my article and — well, I just thought I'd mention it. —I.A.

Murphy's Laws of Technology

The cartoon character of Murphy was an incompetent mechanic at Cape Canaveral around whom the law was formulated: '[When Murphy is around] everything that can go wrong sooner or later will go wrong.' The New York publishing company Harvey Hutter has compiled further such laws:

If several things can go wrong, the one that will cause the most damage will be the one to go wrong.

Any instrument when dropped will roll into the least accessible corner.

Any circuit design must contain at least one part which is obsolete, two parts which are unobtainable and three parts which are still under development.

A failure will not appear until a unit has passed final inspection.

If builders built buildings the way computer programmers write programs, then the first woodpecker that came along would destroy civilization.

Logic is a systematic method of reaching the wrong conclusion with confidence.

An expert is one who knows more and more about less and less until he knows absolutely everything about nothing.

Tell a man there are 100 billion stars in the Galaxy and he'll believe you. Tell him a bench has wet paint on it and he'll have to touch to be sure.

All great discoveries are made by mistake.

Nothing ever gets built on schedule or within budget.

To err is human, but to really foul things up requires a computer.

Any given computer program, when running, is obsolete.

A computer makes as many mistakes in two seconds as twenty people working twenty years.

To spot the expert, pick the one who predicts the job will take the longest and cost the most.

A complex system that works is invariably found to have evolved from a simple system that works.

Computers are unreliable, but humans are even more so. Any system that depends on human reliability is unreliable.

Under the most rigorously controlled conditions of pressure, temperature, volume, humidity and other variables the organism will do as it damn well pleases.

The only perfect science is hind-sight.

Any simple theory will be worded in the most complicated way.

The degree of technical competence is inversely proportional to the level of management.

Appendix

The March of Knowledge

The growth of science, the life blood of civilization, has never been even. It has often taken the form of two steps forward, one step back. In this no doubt incomplete chronology of discoveries and inventions from the earliest known times to the present day, I have italicized those events that seemed to me to have retarded knowledge rather than advanced it.* They were all too frequent, and there will be many more of them in the future.

3000 BC [approximate] The abacus, the first calculating device, invented in the Tigris-Euphrates valley.

450 BC [approximate] Democritus of Thrace suggests that atoms are the basic components of matter.

360 BC [approximate] *Socrates and Plato teach that the use of science for practical purposes is degrading, and that its only proper function is to exercise the mind.*

312 BC Appius Claudius, Censor of the Roman Republic builds the first all-weather road, from Rome to Brindisi, known as the Appian Way.

300 BC [approximate] *Zosimus of Panopolis claims that base metals can be turned into gold, wasting hundreds of thousands of research man-years and setting back the progress of chemistry by nearly two millennia.*

Euclid, in his 'Elements,' systemizes the laws of arithmetic.

250 BC [approximate] Aristarchus of Samos calculates the circumference of the Earth at 25,000 miles.

---

* Wars, curiously enough, usually advance knowledge rather than retard it, because of the military inventions that are afterwards put to civilian use.

130 BC [approximate] *Hipparchus of Nicaea describes the constellations of the Zodiac and pronounces their influence on human fate, thus establishing the pseudoscience of astrology.*

140 AD [approximate] Claudius Ptolemy publishes the first comprehensive map of the stars, known later as the *Almagest*, meaning the 'greatest'.

250 [approximate] Diophantus of Alexandria produces the first book on algebra.

285 Pappus of Alexandria describes five machines in general use: the cogwheel, the lever, the pulley, the screw and the wedge.

325 *The Council of Nicaea, convened by the emperor Constantine, estimates that the world was created 1,384 years before the birth of Abraham.*

391 *The great library of Alexandria destroyed by the Christians during a Roman civil war.*

815 [approximate] The Arabs establish the modern system of numerals, including the zero, making it easy to multiply by 10.

963 Al Sufi, in his 'Book of the Fixed Stars,' refers to the 'nebulae,' which centuries later are shown to be other galaxies beyond our own.

1250 [approximate] *Roger Bacon's books proposing scientific experiments and the use of mathematics are condemned by the Church.*

1252 *Pope Innocent IV introduces the Inquisition for the 'extirpation' of heretics. This severely retards the growth of science in many countries and for many centuries.*

1271 Marco Polo begins his journeys in China.

1303 First recorded use of spectacles.

1313 Gunpowder invented by Berthold Schwartz, a German friar.

1337 William Merlee of Oxford makes the first scientific weather forecasts.

1418 Prince Henry the Navigator of Portugal establishes his school for navigation at Sagres, on Cape St Vincent, enabling Portuguese seamen to explore the coast of Africa, down to the Cape of Good Hope and beyond to the East.

1450   [approximate] *The Government of China prohibits the construction of all ships with more than two masts, thus destroying China as a sea-faring nation.*

1454   Johannes Gutenberg builds the first printing press.

1492   Christopher Columbus reaches the Americas.

1497   The Cabots, father and son, reach the east coast of North America.

1498   Vasco da Gama discovers a sea route to India.

1510   Leonardo da Vinci works out the principle of the water turbine.

1513   Nuñez de Balboa discovers the Pacific Ocean.

1522   The survivors of Magellan's expedition complete the first circumnavigation of the globe.

1527   Paracelsus lectures on medicine at the University of Basel and later produces a manual of surgery.

1541   Gonzalo Pizarro is the first European to explore the Amazon.

1543   Nicholas Copernicus publishes his book *On Revolutions*, showing that the Solar System revolves around the Sun. *But the influential Martin Luther declares Copernicus's discovery is 'anti-Biblical and intolerable'.*

1547   *Nostradamus starts to make bogus predictions about the future in the form of mystical verses, which some people take seriously even today.*

1550   Rheticus publishes trigonometric tables.

1556   Georg Agricola postumously publishes a study of mineralogy.

1572   Tycho Brahe collects astronomical observations that lead to Johannes Kepler's laws of planetary motion.

1595   Geradus Mercator publishes posthumously his global map.

1600   *Giordano Bruno is burned at the stake for proclaiming the existence of an infinite number of planets.*

1608   Johann Lippershey of Holland invents the telescope.

1609   Galileo first looks at the night sky through one of Lippershey's telescopes. He finds the four largest moons of Jupiter, and mountain ranges on our own Moon.
        Kepler publishes his laws of planetary motion.

1611   *The King James Bible contains a statement by Bishop James Ussher of Armagh that the world was created on October 23, 4004 BC.*

1614   John Napier invents logarithms.

1616   *The Roman Catholic Church bans all books that maintain that the Earth moves.*

1620   Francis Bacon publishes his *Great Instauration of Learning* that analyzes scientific method and launches the growth of modern science.

1628   William Harvey describes the circulation of the blood.

1654   Blaise Pascal and Pierre de Fermat develop the theory of probability.

1658   Robert Hooke invents the balance spring for watches.

1660   Charles II founds the Royal Society of London.

1665   Giovanni Cassini determines the rotations of Jupiter, Mars and Venus.

1671   *Gottfried Leibniz insists upon the existence of the ether, which is not disproved until the Michelson-Morley experiment of 1887.*

1675   Olaus Romer makes the first roughly accurate estimate of the speed of light.

1687   Isaac Newton publishes his 'Principia' in Latin, which includes (but does not explain very clearly) his three laws of gravitation. *It is not translated into English until 1729.*

1705   Edmund Halley discovers the true nature of comets.

1710   Jakob Le Blon invents three-colour printing.

1742   Anders Celsius invents the Centigrade scale of temperature.

1751   Denis Diderot and Jean d'Alembert publish the first encyclopedia.

1752   Benjamin Franklin invents the lightning conductor.

1761   Mikhail Lomonosov discovers that Venus has an atmosphere.

1766   Henry Cavendish identifies hydrogen, the most abundant element in the universe.

1772   Daniel Rutherford and Joseph Priestley independently discover nitrogen.

1775   James Watt perfects his invention of the steam engine.

1777   Antonine Lavoisier, the father of modern chemistry, proves that air consists mostly of oxygen and nitrogen. (In announcing his discovery of oxygen, he fails to mention the help he received from Joseph Priestley).

1781   William Herschel discovers the planet Uranus.

1783	The Montgolfier brothers make the first flight, in a balloon at Annonay, France.
1793	Eli Whitney invents the cotton gin, which later helps to precipitate the American Civil War.
1796	Edward Jenner discovers vaccination for smallpox.
1801	Robert Fulton, inventor of the steamship, builds the first submarine. Joseph Lalande catalogues 47,390 stars.
1802	John Dalton introduces atomic theory into chemistry. William Hershel discovers binary stars.
1815	Richard Trevithick builds a steam-powered train. Humphry Davy invents the miners' safety lamp.
1816	Rene Laennec invents the stethoscope.
1818	Jons Berzelius publishes the molecular weights of 2,000 chemical compounds.
1819	David Napier constructs the flat-bed cylinder printing press.
1823	Charles Babbage starts his efforts to build a mechanical computer.
1824	Pierre Prevost and Jean Dumas prove that the sperm is essential to fertilization in mammals.
1825	Opening of the first passenger-carrying railway, from Stockton to Darlington.
1831	Michael Faraday demonstrates an electric generator.
1837	Samuel Morse demonstrates his electric telegraph.
1839	Louis Daguerre develops photography. Charles Goodyear develops the commercial use of rubber.
1840	Justus von Leibig invents artificial fertilizer.
1842	Crawford Long performs the first surgical operation using anasthesia.
1846	Urbain Leverrier and J.C. Adams independantly discover the planet Neptune.
1850	W.C. Bond takes the first astronomical photograph.
1853	Samuel Colt revolutionizes the manufacture of handguns.
1859	Charles Darwin proposes the theory of evolution in *The Origin of Species*.
1861	Discovery of a fossil showing the evolutionary link between reptile and bird.
1864	Julius Sachs demonstrates photosynthesis.
1866	Alfred Nobel invents dynamite. Gregor Mendel establishes the science of genetics.

1867   Pierre Michaux begins to manufacture bicycles.

1868   Discovery of skeletons of Cro-Magnon and Neanderthal men.

1869   Dimitri Mendeleev develops the periodic table of the chemical elements, predicting the properties of elements not yet discovered.

1870   Louis Pasteur and Robert Koch establish the germ theory of disease. Pasteur develops vaccines for anthrax and rabies.

1871   Simon Ingersoll invents the pneumatic rock drill. It is used to open the Mount Cenis Tunnel.

1873   James Clerk Maxwell discovers that electricity and magnetism are different manifestations of the same fundamental force.

1874   Lord Kelvin pronounces the Second Law of Thermodynamics.

1876   Alexander Graham Bell invents the telephone.

1879–80 William Herschel* and Henry Faulds independently discover the uniqueness of fingerprints.

1880   Thomas Edison and Joseph Swann independently invent electric lighting.

1882   Albert Michelson measures the speed of light at 186,000 miles per second.

1887   Michelson and Edward Morley conduct their famous experiment showing that the ether wind does not exist, paving the way for Einstein's Special Theory of Relativity. H.W. Goodwin invents culluloid film.

1888   Heinrich Hertz discovers radio waves.
        John Dunlop invents the pneumatic tyre.

1893   Karl Benz and Henry Ford build their first cars.

1894   Louis Lumiere invents the cinematograph.
        Emil Berliner builds the first gramophone.

1895   Wilhelm Rontgen identifies X-rays.

1896   Antoine Becquerel determines the existence of radioactivity.

1897   Martinus Beijerinck discovers viruses.

1898   Sir Harold Spencer Jones finds Eros, the largest of the asteroids, or minor planets.

1901   The first motor-cycles

* No relation to the famous astronomer.

1903   Wilbur and Orville Wright make the first powered flight, at Kitty Hawk, North Carolina.

1905   Albert Einstein publishes his Special Theory of Relativity, which shows that no material object can attain the speed of light.

1906   R.A. Fessenden makes the first radio broadcast.

1909   Paul Ehrlich prepares Salvarsan for the cure of syphilis.
        Leo Baekeland develops Bakelite, the first of the plastics.

1911   Charles Kettering invents the first practical electric self-starter for cars.
        Heike Kamerlingh-Onnes discovers superconductivity.

1913   Henry Norris Russell and Einar Hertzsprung plot the brightness and colours of stars, making possible the Hertzsprung-Russell Diagram which explains the evolution of stars from birth to death.

1914   Walter Adams and Arnold Kohlschutter determine the absolute luminosity of stars from their spectra alone, making it possible to determine the distance of millions of distant stars.

1916   Einstein publishes his General Theory of Relativity, introducing the idea of gravity being caused by curved space-time. It indirectly predicts the existence of black holes.

1918   Harlow Shapley determines the shape of our Milky Way galaxy.

1919   Einstein's General Theory is proved correct during an eclipse of the Sun.
        Lenin says in a letter to Maxim Gorky: 'Scientists imagine that they are the brains of the nation. Actually, they are not the brains but the shit.' Until the coming of the atomic bomb, Soviet scientists are treated as such.

1921   Hermann Oberth writes his dissertation 'The Rocket into Interplanetary Space.'

1924   Edwin Hubble discovers the existence of billions of galaxies beyond the Milky Way.
        Arthur Eddington discovers that the brightness of a star is related to its mass.

1925   *Trofim Lysenko proclaims that the acquired characteristics of wheat (or any other species) can be inherited. With Stalin's help, he persecutes dissenting biologists. Soviet genetics and biology are set back a quarter of a century.*
        John Logie Baird demonstrates television.

John Scopes, a Tennessee schoolteacher, is put on trial for teaching Darwin's theory of evolution.

1926   Robert Goddard fires a liquid-fuelled rocket, a precursor of the Saturn 5 Moon rockets and the space shuttle.

Hubble discovers that the universe is expanding.

1928   Alexander Fleming discovers penicillin.

1930   Clive Tombaugh discovers the planet Pluto.

1935   Karl Jansky discovers radio waves from outer space, starting the science of radio astronomy.

James Chadwick discovers the neutron.

Robert Watson-Watt develops radar.

1937   Niels Bohr and John A. Wheeler develop the theory of nuclear fusion.

Frank Whittle constructs the first jet engine.

1938   Lajos Biro invents the ballpoint pen.

1939   Igor Sikorsky builds the first helicopter.

J. Robert Oppenheimer and Hartland Snyder predict the existence of black holes.

1941   Donald Bailey invents the portable bridge.

1942   Enrico Fermi, in a Chicago squash court, achieves the first self-sustaining nuclear fission reaction.

John von Neumann builds the first computer.

Invention of magnetic tape for storing data.

1943   Jacques-Yves Cousteau demonstrates the aqualung.

1945   The first atomic bomb is detonated.

Arthur C. Clarke proposes stationary satellites for communication.

1947   Charles 'Chuck' Yeager breaks the sound barrier.

Thor Heyerdahl sails on a raft from Polynesia to Peru in 101 days to prove prehistoric migration.

Invention of the transistor.

1948   John von Neumann works out the mathematics of self-reproducing machines.

The 200-inch Mount Palomar telescope goes into service.

The first long-playing gramophone record.

1950   Alan Turing proposes the 'Turing test' to test the intelligence of computers.

1951   The first electric power from atomic energy.

1952   First hydrogen bomb exploded.

1953   James Watson and Francis Crick discover the molecular

	structure of DNA, the core of all life.
1957	Russia's Sputnik 1, the first man-made satellite is launched into Earth orbit.
1958	US nuclear submarine *Nautilus* sails under the North Polar icecap.
1960	The first weather satellite, Tiros I, transmits television pictures of cloud cover around the world.
1961	Yuri Gagarin is the first man in space. Carl Sagan proposes the 'terra-forming' of Venus to make it fit for human habitation. President John F. Kennedy commits the United States to landing a man on the Moon before 1970.
1962	Herman Kahn introduces 'flexible response' into international disputes to reduce the risk of nuclear war.
1963	T.A. Matthews and Alan Sandage discover quasars. Michael de Bakey makes the first use of an artificial heart in a patient who survives.
1965	Edward White makes the first space walk.
1969	Neil Armstrong and Edwin 'Buzz' Aldrin walk on the Moon.
1972	President Richard Nixon orders the construction of space shuttles.
1973	*Cuts by the U.S. Congress force the abandonment of the first phase of manned lunar exploration.*
1976	Viking spacecraft soft-lands on Mars and finds no traces of life.
1981	The first shuttle flight into space.
1983	President Ronald Reagan orders the construction of a permanent manned space station.
1986	*The space shuttle Challenger explodes on its way to orbit, setting back American manned space travel by nearly three years.* *A nuclear reactor explosion at Chernobyl in Russia undermines world faith in nuclear power.* Europe's Giotto spacecraft penetrates Halley's Comet.
1987	The supernova explosion 1987A, in the Greater Magellanic Cloud, emits traces of iron and nickel, proving that the heavier elements in our bodies came from exploding stars. The European Space Agency decides to construct Ariane-5, the first manned space rocket not built by the superpowers.

1988   Discovery in space of a pulsar that rotates like a clock which only gains or loses a second in three million years.

Successful flight by the shuttle Discovery restores confidence in Western manned space flight.

1989   The spacecraft Voyager 2 flies by Neptune, 2,700 million miles from Earth, and discovers eight moons and three rings.

CERN's Large Electron Positron Collider, a particle accelerator with a circumference of 16.8 miles, comes into operation.

The first trans-oceanic optical fibre cable, capable of carrying 40,000 simultaneous telephone conversations, is laid between Europe and the United States.

1990   The Hubble Space Telescope is launched into orbit, *albeit with a flawed mirror.*

Fiat of Italy and Peugeot of France put on sale the first electric passenger cars.

Launch into orbit of ultraviolet observatories ROSAT and Astro-1.

Jean-Marie Lehn, Ulrich Koert and Margaret Harding report the synthesis of a new class of compounds, called nucleohelicates, that mimic the double-helical structure of DNA, turned inside out.

1991   The Galileo space probe flies past the asteroid Gaspra, approaching it to within 16,000 miles.

The first satellite car navigation system goes on sale, in Japan.

A borehole in the Kola Peninsula in Arctic Russia, begun in the 1970s, reaches a depth of 40,240 feet where the temperature is 410°F (210°C) It is expected to reach 49,000 feet by 1995.

ICI begins production of the hydrofluorocarbon HFA-134a, a substitute for CFCs in refrigerators and air-conditioning systems, aimed at protecting the Earth's ozone layer.

The first controlled production of nuclear fusion energy, at the Joint European Torus (JET), at Culham, Oxfordshire.

1992   George Smoot, using the Cosmic Background Observer satellite (COBE), detects the ripples of the Big Bang that started the universe some 15,000 million years ago.

The world's largest plant, a fungus covering 1,480 acres, is discovered in Washington state, USA.

Transistors made from superconducting ceramics rather than semiconductors produced in Japan by Sanyo Electric. They are 10 times faster than semiconductor transistors.

1993   Andrew Wiles proves Fermat's Last Theorem, one of the most baffling challenges in pure mathematics.

NASA loses contact with Mars Observer as the spacecraft goes into orbit around Mars.

Bibliography

(In case of two more authors or editors, the book is listed under whichever is first in alphabetical order.)

ALLEN, C.W.: *Astrophysical Quantities* (Athlone Press, London, 1973).

ALTMAN, Lawrence K: *Who Goes First: The Story of Self-Experimentation in Medicine* (Random House, New York, 1987).

ASIMOV, Isaac: *Asimov's Biographical Encyclopedia of Science and Technology* (Doubleday, New York, 1964).
_____ *Only a Trillion* (Grosset and Dunlap, New York, 1976).

BAKER, David: *The History of Manned Spaceflight* (New Cavendish Books, London, 1981).

BARROW, John: *The World within the World* (Clarendon Press, Oxford, 1988).

BELL, E.T.: *Men of Mathematics* (Gollancz, London, 1937).

BERGAMINI, David. With Henry Margenau: *The Scientist.* (Life Science Library, 1964).
_____ *Mathematics* (Life Science Library, New York, 1963).

BERGIER, Jacques. With Louis Pauwels: *The Morning of the Magicians* (Mayflower, London, 1971. Originally published in France with the title *The Dawn of Magic*).

BERRY, Adrian: *The Next Ten Thousand Years: A Vision of Man's Future in the Universe* (Cape, London, 1974).
_____ *The Super-Intelligent Machine: An Electronic Odyssey* (Cape, London, 1983).

BOND, Peter: *Heroes in Space: From Gagarin to Challenger* (Basil Blackwell, Oxford, 1987).

BOVA, Ben: Introduction to *Escape Plus*, volume of short science fiction stories (Methuen, London, 1988).

BRADFORD, Ernle: *Southward the Caravels: The Story of Henry the Navigator* (Hutchinson, London, 1961).

BRONOWSKI, Jacob: *The Ascent of Man* (British Broadcasting Corporation, London, 1974).

BRUCE, Robert V.: *Alexander Graham Bell and the Conquest of Solitude* (Gollancz, London, 1973).

CARSWELL, John: Introduction to *The Singular Travels, campaigns and Adventures of Baron Munchausen*, by R.E. Raspe and others (The Cresset Press, London, 1948).

CARVER, Larry. With Hans Mark: 'Challenger' and Chernobyl. *Interdisciplinary Science Reviews* (Vol. 12, No,3, Sept. 1987).

CHORLTON, Windsor. With the Editors of Time-Life Books: *Ice Ages* (Time-Life, Amsterdam, 1983).

CHURCHILL, Winston S.: *Thoughts and Adventures* (Thirnton Butterworth, London, 1932).

CLARK, Ronald W.: *The Survival of Charles Darwin: A Biography of a Man and an Idea* (Weidenfeld, London, 1984).

CLARKE, Arthur C.: *Voices from the Sky: A Preview of the Coming Space Age* (Harper and Row, New York, 1965).
_____ *Ascent to Orbit: A Scientific Autobiography: The Technical Writings of Arthur C. Clarke* (John Wiley, New York, 1984).

COCHRANE, J.A.: *Lavoisier* (Constable, London, 1931).

COUSTEAU, J.Y. With Frederic Dumas: *The Silent World* (Hamish Hamilton, London, 1983).

CRITCHFIELD Margot. With Thomas A. Dwyer: *A Bit of IBM Basic* (Addison-Wesley, Reading, Massachusetts, 1984).

CROMBIE, A.C.: *Augustine to Galileo: The History of Science, AD 400–1650* (Harvard University Press, 1953).

CRONIN, Vincent: *The View from Planet Earth: Man Looks at the Cosmos* (Collins, London, 1981).

DE KRUIF, Paul: *Microbe Hunters* (Harcourt, Brace, New York, 1926).

DOYLE, Sir Arthur Conan: *The Conan Doyle Historical Romances* (London, 1931).

DYSON, Freeman: *Infinite in All Directions* (Harper and Row, New York, 1988).

EDEY, Maitland A. With Donald C. Johanson: *Lucy: The Beginnings of Mankind* (Granada, London, 1981).

EINSTEIN, Albert. With Hendrik Lorentz, Herman Weyl and Hermann Minkowski: *The Principle of Relativity: A Collection of Original Memoirs on the Special and General Theories of Relativity* (Dover Publications, New York, 1923).

EISELEY, Loren: Biographical essay on Charles Darwin. *Scientific Genius and Creativity (Scientific American* reprint. W.H. Freeman, New York, 1982).

FALLACI, Oriana: *If the Sun Dies* (Collins, London, 1967).

FARRINGTON, Benjamin: *Francis Bacon: Philosopher of Industrial Science* (Macmillan, London, 1973).

FAULDS, Henry: 'On the Skin Furrows of the Hand' (*Nature*, Oct. 28, 1880).

FERRIS, Timothy: *Coming of Age in the Milky Way* (William Morrow, New York, 1988).

FEYNMAN, Richard P.: *'Surely You're Joking, Mr Feynman!': Adventures of a Curious Character* (W.W. Norton, New York, 1985).

FLANAGAN, Dennis: *Flanagan's View: A Spectator's Guide to Science on the Eve of the 21st Century* (Alfred A. Knopf, New York, 1988).

FOSTER, Caxton: *Cryptanalysis for Microcomputers* (Hayden Books, Rochelle Park, New Jersey, 1982).

FRENCH, Sidney J.: *Torch and Crucible: The Life and Death of Antoine Lavoisier* (Princeton University Press, 1941).

FURNEAUX, Robin: *The Amazon: The Story of a Great River* (Hamish Hamilton, London, 1969).

GARDNER, Martin: *Relativity for the Million* (Macmillan, New York, 1962).

GIBBON, Edward: *The Decline and Fall of the Roman Empire.* (London, 1862).

GOOD, I.J.: (Ed.) *The Scientist Speculates: An Anthology of Partly-Baked Ideas* (Heinemann, London, 1962).*

GRUN, Bernard: *The Timetables of History: A Chronology of World Events from 5,000 BC to the Present Day* (Thames and Hudson, London, 1975).

HALSTEAD, Beverley: *The Hard and Woolly Sciences* (Address to the British Association for the Advancement of Science, 1986).

HAWKING, Stephen: *A Brief History of Time: From the Big Bang to Black Holes* (Bantam Press, London, 1988).

HOGAN, James P.: *Endgame Enigma* (Century Hutchinson, London, 1988).

HOGBEN, Lancelot: *Mathematics for the Million* (Allen and Unwin, London, 1936).

HOSS, Norman. With William B. Sill: *The New Popular Encyclopedia of the Sciences* (Allen and Unwin, London, 1963).

HOWARD, Peter: 'The Dangerous Deserts of Space' (*Discovery*, June 1959).

INNES, Mary M.: *Ovid's Metamorphoses* (Penguin Classics, London, 1955).

---

* The first eight definitions in the 'Glossary of Incompetence' first appeared in the journal *Shipbuilding and Shipping Record*, Vol. 91, p. 301, 1958. The next twelve were first published in *Metal Progress*, Vol. 71, pp. 75-76, 1957. The last two were apparently created by Good himself.

IVANOVIC, Vane: *LX: Memoirs of a Yugoslav* (Weidenfeld, London, 1977).

JACOB, Henry Edward: *The Saga of Coffee: The Biography of an Economic Product* (Allen and Unwin, London, 1935).

JAFFE, Bernard: *Michelson and the Speed of Light* (Heinemann, London, 1961).

JOHANSON, Donald. *See* Edey.

JUNGK, Robert: *Brighter than a Thousand Suns: The Moral and Political History of the Atomic Scientists* (Gollancz, in association with Rupert Hart-Davis, London, 1958).

KAHN, David: *The Codebreakers: The Story of Secret Writing* (Weidenfeld and Nicolson, London, 1966).

KESTEN, Hermann: *Copernicus and his World* (Secker and Warburg, London, 1945).

KUHN, Thomas S.: *The Structure of Scientific Revolutions*. (University of Chicago Press, 1965).

LAMONT, Lancing: *Day of Trinity* (Hutchinson, London, 1966).

LANDSTROM, Bjorn: *The Ship: A Survey of the History of the Ship from the Primitive Raft to the Nuclear-Powered Submarine, in Words and Pictures* (Allen and Unwin, London, 1961).

LAPP, Ralph E.: *Matter* (Life Science Library, New York, 1965).

LEY, Willey: *The Poles* (Life Nature Library, New York, 1966).

MACAULAY, Thomas Babington: *Lord Bacon* (London, 1852).

— *History of England,* Vol. 1, Chap. 3 (London, 1866).

McKAY, Alwyn: *The Making of the Atomic Age* (Oxford University Press, 1984).

McKIE, Douglas: *Antoine Lavoisier: Scientist, Economist, Social Reformer* (Constable, London, 1952).

MEDAWAR, Peter and Jean: *Aristotle to Zoos: A Philosophical Dictionary of Biology* (Oxford University Press, 1985).

MICHAELIS, Anthony R.: *From Semaphore to Satellite* (The International Telecommunication Union, Geneva, 1965).

MICHELMORE, Peter: *Einstein: Profile of the Man* (Frederick Muller, London, 1963).

MOORE, Dan Tyler. With Martha Waller: *Cloak and Cipher* (Harrap, London, 1964).

MOORE, Patrick: *Patrick Moore's History of Astronomy.* (Macdonald, London, 1961).

PEARSON, Jerome: 'The Lonely Life of a Double Planet' (*New Scientist*, August 25, 1988).

PIGAFETTA, Antonio: *Magellan's Voyage: A Narrative Account of the First Circumnavigation* (Translated and edited by R.A. Skelton from the manuscript in the Beinecke Rare Book and Manuscript Library of Yale University. Yale University Press, New Haven and London, 1969).

PLUTARCH: *Lives of Illustrious Men.* Translated from the Greek by J. and W. Langhorne (London, 1866).

PORTER, Roy (Ed.): *Man Masters Nature: 25 Centuries of Science* (BBC Books, London, 1987).

PRESCOTT, William H.: *History of the Conquest of Peru.* (London, 1847).

RANDI, James: *The Faith Healers* (Prometheus Books, Buffalo, New York, 1987).

RASPE, R.E. *See* Carswell, John.

REGIS, Ed.: *Who Got Einstein's Office? Eccentricity and Genius at the Institute for Advanced Study* (Addison-Wesley, Reading, Massachusetts, 1987).

RONAN, Colin: *The Cambridge Illustrated History of the World's Science.* (Cambridge University Press, 1983).

RUSSELL, Bertrand. *The ABC of Relativity* (Allen and Unwin, London, 1925).

SEELIG, Carl: *Albert Einstein: A Documentary Biography* (Staples Press, London, 1956).

SOCRATES SCHOLASTICUS: *A History of the Church from Constantine to Theodosius* (Translated from the Greek. London, 1841).

SWETZ, Frank J.: *Capitalism and Arithmetic: The New Math of the 15th Century* (Open Court Press, La Salle, Illinois, 1988).

THORWALD, Jurgen: *The Marks of Cain: The Century of the Detective* (Thames and Hudson, London, 1965).

VITRUVIUS POLLIO: *De Architectura* (Translated by F. Krohn and M.H. Morgan. London, 1914).

WEBER, R.L.: *A Random Walk in Science* (The Institute of Physics, London, 1973).

YEAGER, Chuck. With Leo Janos: *Yeager: An Autobiography* (Century Hutchinson, London, 1986).

WEIG, Stefan: *Magellan: Pioneer of the Pacific* (Cassell, London, 1938).

— *The Queen of Scots* (Cassell, London, 1935).

Quoted material: acknowledgments

Acknowledgement is made to the following sources for the quoted material.

Every effort has been made to trace the copyright holders, and the author and publisher would like to apologize to anyone whose copyright may have been infringed.

The page numbers printed here refer to this book.

Ernle Bradford: *Southward the Caravels: The Story of Henry the Navigator* (Hutchinson, London, 1961). The author thanks the estate of Ernle Bradford for permission to publish this extract.   17–20

Viscount Furneaux: *The Amazon: The Story of a Great River* (Hamish Hamilton, London, 1969).   26–28 28–30

Windsor Chorlton: *Planet Earth 'Ice Ages'* (Time-Life Books Inc., 1984).   30–33

Willy Ley and the Editors of Time-Life Books. From LIFE NATURE LIBRARY: *The Poles* (copyright 1977 Time-Life Books Inc.).   33–35

J. Y. Cousteau, with Frederic Dumas: *The Silent World* (Hamish Hamilton, London, 1983).   36–37

Mary M. Innes: *Ovid's Metamorphoses* (Penguin Classics, London, 1955).   38–40

Peter Howard: 'The Dangerous Deserts of Space'. Article in *Discovery*, 1959.   40–44

Chuck Yeager, with Leo Janos: *Yeager: An Autobiography* (Century Hutchinson, London, 1986).   44–45

Oriana Fallaci: *If The Sun Dies* (Collins, London, 1967).   45–46

Daily Telegraph (Syndication Department), 1 Canada Square, Canary Wharf, London E14: Report by Adrian Berry, 22 July 1969, of Armstrong and Aldrin walking on the Moon.   46–49

Peter Bond: *Heroes in Space: From Gagarin to Challenger* (Basil Blackwell, Oxford, 1987). The author has used an edited version.   49–56

Dr Hans Mark and Professor Larry Carver: 'Challenger and Chernobyl' (a paper first published in *Interdisciplinary Science*

Reviews, 12, 241–252, 1987 and reprinted here by permission of J. W. Arrowsmith Ltd, without the original accompanying illustrations).   56–63

Patrick Moore: *Patrick Moore's History of Astronomy* (Macdonald, London, 1961).   68–71

Stephen Hawking: from *A Brief History of Time: From the Big Bang to Black Holes* (published by Bantam Press, London, 1988, copyright Space Time Publications 1988).   71–72

Martin Gardner: *Relativity for the Million* (Macmillan, New York, 1962).   75–78

Jerome Pearson: 'The Lonely Life of a Double Planet' (*New Scientist*, 25 August 1988).   82–86

Timothy Ferris: *Coming of Age in the Milky Way* (William Morrow, New York, 1988).   86–90

Dr Anthony R. Michaelis: *From Semaphore to Satellite* (copyright International Telecommunication Union, Geneva, 1965, who authorized permission to reproduce pages 109 to 112 of this book).   91–94

I.J. Good (ed.) from *An Anthology of Partly-Baked ideas* (William Heinemann Ltd., London, 1982. Reprinted by permission of the publisher).   96–98

David Kahn: excerpts from *The Codebreakers: The Story of Secret Writing* (Weidenfeld and Nicolson). Copyright (New York and London) 1967, 1968.   99–101
                                                                 103–105

Margot Critchfield, with Thomas A. Dwyer: *A Bit of IBM Basic* (Addison-Wesley, Reading, Massachusetts, 1984).   105–108

Ed Regis: *Who Got Einstein's Office? Eccentricity and Genius at the Institute for Advanced Study* (Addison-Wesley, Reading, Mass, USA, 1987).   110–112

Alan Turing: 'Computing Machinery and Intelligence' in *Mind: A Quarterly Review of Psychology and Physiology,* Vol. 59, Oct. 1950.   113–119

Alan Moorehead: *Darwin and the Beagle* (Hamish Hamilton, London, 1969).   122–127

Ronald W. Clark: *The Survival of Charles Darwin: A Biography of a Man and an Idea* (Weidenfeld & Nicolson, London, 1984).   127–128

Maitland Edey, with Donald Johanson: *Lucy: The Beginnings of Mankind* (Grafton Books, London – formerly Granada – 1981).   128–134

Dr. L. B. Halstead: 'The Hard and Woolly Sciences'; an address to the British Association for the Advancement of Science, given in Bristol, 1986, in the session 'Science and Pseudoscience' chaired by Dr Helen Haste.   135-138

J. A. Cochrane: *Lavoisier* (Constable, London, 1931).   143-145

Eric Temple Bell: Men of Mathematics: *The Story of Evariste Galois* (Gollancz, London, 1937). Copyright 1937 E. T. Bell, renewed 1965 by Taine T. Bell.   145-151

Lawrence Altman: *Who Goes First?* Copyright 1986, 1987 by Lawrence Altman. Reprinted by permission of Random House, Inc.   151-155 201-202

Peter and Jean Medawar: *Aristotle to Zoos: A Philosophical Dictionary of Biology* (Weidenfeld and Nicolson, London, 1985).   156-157

Jacques Bergier, with Louis Pauwels: *The Morning of the Magicians* (Grafton Books, London, 1971).   157-158

Arthur C. Clarke: *Voices from the Sky: A Preview of the Coming Space Age* (Gollanz, London, 1974).   159-163

James Randi: reprinted from *The Faith Healers*, with permission of Prometheus Books, Buffalo, New York State. Published 1987).   164-166

Henry Edward Jacob: *The Saga of Coffee: the Biography of an Economic Product* (Allen and Unwin, London, 1935).   167-169

Frank Swetz: reprinted from *Capitalism and Arithmetic* by permission of The Open Court Publishing Company, La Salle, Illinois. Published 1988).   170-172

Isaac Asimov: *Only a Trillion* (T. Y. Crowell, New York, 1957).   173-174 218-225

Peter Michelmore: *Einstein: Profile of the Man* (Frederick Muller, London, 1963).   181-183 185-186

Henry Margenau, David Bergamini and the Editors of Time-Life Books. From LIFE SCIENCE LIBRARY: *The Scientist* (copyright 1971 Time-Life Books Inc.).   184

Lancing Lamont: *Day of Trinity* (Hutchinson, London, 1966).   186-187

Ben Bova: Introduction to *Escape Plus*. Reproduced by permission of Methuen, London, and the Barbara Bova Literary Agency, Hertford, Connecticut. Published 1988.   189-191

Vincent Cronin: *The View from Planet Earth* (Collins, London, 1981).   199-201

Paul de Krief: *Microbe Hunters* (Harcourt Brace Jovanovich, Inc., New York, 1926).   203-206

Richard P. Feynman: *Surely You're Joking, Mr Feynman! Adventures of a Curious Character* (W. W. Norton, New York, 1985). These are excerpts from Part 5, in correct sequence, but are quoted with certain omissions which do not affect the overall sense.   206–211

Vane Ivanovic: *Memoirs of a Yugoslav* (Weidenfeld & Nicolson, London, 1977).   211–213

John Carswell: Introduction to *The Singular Travels, Campaigns and Adventures of Baron Munchausen*, by R. E. Raspe and others (Cresset Press, London, 1948).   214–217

Index

Abacus, 228
Accident, 56, 60-3
Accounting, 170-2
Aeronautics: *Apollo 13* mission, 49-56; *Challenger* explosion, 56-8, 60-3; early balloon flight, 40-4; first Moon landing, 46-9; sound barrier broken, 44-5
Africa, 18, 20
Agricola, Georg, 230
Agriculture, 199
Alamagordo explosion, 186-7
Alchemy, 71, 199, 228
Aldrin, Colonel Edwin 'Buzz', 46-9, 236
Alexandria, 139-40, 229
Algebra, 145-8, 150-1, 171, 229
Alphabet, 98-9
Alphonso X, 68
Altman, Lawrence, 151, 201, 202
Alvarez, Luis, 136-7
Amazon River, 23-8
'Amazons', 26-8
America, discovery of, 230
Amundsen, Roald, 33, 34
Anaesthesia, 232
Analytical Engine, 118
Anatomy, 153-4
Andromeda galaxy, 90
Anoxia, 41-4
Antarctic, 33-5
Apollo 9, 52
Apollo 11, 46-9
Apollo 13, 49-56
Appian Way, 228
Aqualung, 35-7, 235
Aquarius lunar module, 49-56
Archimedes, 175-9
Ariane-5, 236
Aristarchus of Samos, 64
Aristotle, 35, 69, 70, 140, 156-7
Arithmetic, 170-2, 228
Armstrong, Neil, 46-9, 236
ASCII code, 102, 104
Asimov, Isaac, 173-4, 190, 218-26
Asteroids, 84, 136, 233, 237
Astro-1, 237
Astrolabe, 20
Astrology, 70, 72, 199, 200-1, 229
Astronomy, 139, 229, 230, 231; Copernicus' theory, 64-7, 70; inter-galactic communications, 86-90; planetary tables, 68; solar system, 83-5; space travel, 75-80, 82; supernova,

69; Tycho Brahe's discoveries, 68-70
Atlantic Ocean, 16-8, 19, 21, 26
Atomic bomb, 181-8, 235
Atoms, 182-3, 218-25, 228

Babbage, Charles, 118, 232
Bacon, Francis, 7, 156, 191-7
Bacon, Roger, 196, 229
Bailey bridge, 235
Baird, John Logie, 234
Bakelite, 234
Balboa, Nuñez de, 230
Balloon flights, 40-4, 232
Bar, Reymers, 70
Barometers, 198
Basilisk, 195
Beagle, HMS, 122-7
Becquerel, Antoine, 233
Behaim, Peter, 21, 22
Bell, Alexander Graham, 94, 233
Bell, Eric Temple, 145
Bell Co., 96
Benkendorf, 30-2
Bensalem, 192-7
Bergamini, David 183
Bergier, 157
Bert, Paul 40, 43
Berzelius, Jons, 232
Beta particles, 221, 223, 224, 225
Betancourt, 93
Bicycles, 233
Big Bang, 80-1, 237
Biology, 157
Biro pen, 235
Black holes, 234, 235
Blood, circulation of, 231
Body, human: composition of, 218-25
Bond, Alan, 83
Bond, Peter, 49
Bond, W. C., 232
Borehole, 237
Botany, 199
Bouguer, 29-30
Bouilland, Jean, 95
Bourdillon, Sir James, 109-10
Bova, Ben, 189
Bowers, Lt Henry, 33
Boyle, Robert, 156, 199
Bradbury, Ray, 190
Bradford, Ernle, 17-18
Brahe, Tycho 68-70, 230
Brand, Vance, 54
Braun, Wernher von, 45-6
Bronowski, Jacob, 187-8

Bruno, Giordano, 230
Buller, E. Reginald, 76
Bunsen, Robert, 201-2
Bush, Dr Vannevar, 188

C.M., 92-3
Caesar, Julius, 98
Calcium, 219
Calculus, 72
Calendar, 67
Cancer, 218, 221-2, 225
Cannon, 179-81
Caoutchouc, 28
Caravel, 17-20
Carbon, 219, 222, 223
Carbon-14, 218, 222-6
Carbon dioxide, 83, 222, 223
Carbon monoxide, 42, 59
Car navigation systems, 237
Carrion, Daniel, 151-3
Carswell, John, 214
Carta Pisana, 19
Carvajal, 27
Carver, Professor Larry, 56
Cassini, Giovanni, 231
Catholic Church, 64-7, 139-40, 199-200, 229, 231
Cauchy, Augustin, 147, 148
Cavendish, Henry, 142, 231
Celsius, Anders, 231
Cells, 218-21, 223-5
Ceramics, 238
CERN, 237
CFCs, 237
Chadwick, Sir James, 183, 235
Challenger explosion, 56-8, 60-3
Chappe, Claude, 91-2
Charlatans 164-6
Charles II, 198, 231
Charts — *see* maps
Chemistry, 140-1, 232
Chernobyl disaster, 56, 58-60, 61, 62, 63
Chevalier, Auguste, 151
China, 33, 229, 230
Chlorine, 219
Chlorofluorocarbons (CFCs), 237
Chorlton, Windsor, 30
Christianity, 139-40, 169-70
Chronometer, marine, 173
Cinematograph, 233
Ciphers: one-time, 101-8; substitution, 98-101
Clausius, Rudolf, 73
Clark, Ronald, 127
Clarke, Arthur C., 95, 158-9, 235

Climate, 85
Clocks, 19, 77
Cobalt, 219, 220, 223
Cochrane, J, A., 143
Codes — see ciphers
Coffee, 167-9
Coffinhal, 143-4
Collins, Michael, 49
Colt, Samuel, 232
Columbus, Christopher, 15-16,
  21, 230
Comets, 70, 231, 236
Communication: ciphers,
  98-108, interstellar, 86-90;
  satellites, 235; telephones,
  94-6; visual telegraphs,
  91-2
Compass, magnetic, 19
Computers, 102, 105, 110-19,
  189, 190, 226-7, 232, 235
Computing, 170, 171-2
Conan Doyle, Sir Arthur,
  179-81
Condamine, Charles-Marie de la,
  28-30
Conquistadors, 22-7
Cope, Edward, 135
Copernicus, Nicholas, 13, 64-7,
  70, 230
Copper, 219, 220
Coppleson, V. M., 213
Cosmic Background Observer
  satellite (COBE), 237
Cosmic rays, 222, 223, 225
Cosmic ripples, 237
Cotton gin, 232
Council of Nicaea, 229
Cousteau, Jacques-Yves, 35-7,
  235
Cowley, Abraham, 156
Cresques, Abraham, 19-20
Crick, Francis, 235-6
Critchfield, Martin, 105
Croce-Spinelli, Joseph, 40-3
Cronin, Vincent, 199
Cryptograms, 104
Curies, 59-60
Cyril of Alexandria, 139-40

Daedalus, 38-40
Daguerre, Louis, 232
Dalton, John, 2232
D'Andeli, Henri, 171-2
Dantzig, Tobias, 171
Darwin, Charles, 122-8, 232
Davy, Humphry, 232
Death, survival after, 159
Deep Thought chess program,
  115
Democritus, 228
Denmark, 70-1
Dentistry, 164-5
Descartes, René, 150
Dickinson, Goldsworthy, 156
Dinosaurs, 135-8

Diseases, 151-5, 202, 203-6
Diving, underwater, 35-7
DNA, 235-6, 237
Drake, Frank 82, 83
Dryden, John, 198
Dumas, Jean, 232
Dunlop tyre, 233
Dwyer, Thomas A., 105
Dynamite, 232

Earth: circumference at centre,
  228; circumnavigation of,
  20-2, 230; Copernicus'
  theory, 64-7, 70; 'escape
  velocity' of, 44; expedition to
  Equator to test Newton's
  theory, 28-30; Hitler's theory,
  158; interior of, 33; life on,
  82-6; movement of, 74, 77;
  speed of sound, 44-5;
  temperature of, 218;
  uniqueness of, 83-4; weight
  of atmosphere on surface, 222
Ecuador, 28
Eddington, Sir Arthur, 73
Edison, Thomas, 95, 233
Einstein, Albert, 74-80, 140,
  181-6, 234
El Dorado, 23, 28
Electricity, 93-4, 148, 232, 233
Electrostatic energy, 93-4
Elements, chemical, 141, 233
Elizabeth 1, 99-101
Encyclopedia, 231
Energy, 73, 78, 181, 182
Engines of war, 176-8
Enigma Code, 99
Equations, 145-8, 150-1
Equator, expedition to, 28-30
Eros, 233
Ether, the, 74, 231
Evans, Edgar, 33, 34
Evolution: of man, 132-4; of
  species, 125-8
Extinctions, 32-3, 135-8
Extraterrestrial intelligence,
  82-90, 158, 159, 160, 161

Faber, Felix, 19
Faith-healing, 164-6
Fallaci, Oriana, 45-6
Faraday, Michael, 232
Faulds, Henry, 108
Feathers, 38
Feinberg, Gerald, 77
Fermat, Pierre de, 231; Last
  Theorem, 238
Fermentation, 204-5
Fermi, 173-4
Fermi, Enrico, 78, 79, 173, 183,
  184, 185, 187, 235
Ferris, Timothy, 15-16, 86-90
Fertilizer, artificial, 232
Fessenden, R. A., 234

Feynman, Richard, P., 206-11
Fiat, 237
Fibonacci, Leonardo, 170
Fingerprints, 108-10, 233
Fire, 141, 142
Firedamp, 201
First World War, 162
Fischer, Dr Heinz, 158
Fitzgerald, George, 78
Fitzroy, Captain Robert, 122-8
Flamsteed, John, 72
Flanagan, Dennis, 135, 163
Fleming, Alexander, 235
Flight: human-powered, 38-40;
  sound barrier broken, 44-5
Florentz, Chris, 165
Fluorine, 218, 219
Flying saucers, 159, 160, 162
Flynn, Luke, 84
FORECASTS, 189
Fossils, 128-34, 135-8, 232
France, 146-51, 199-200;
  Academy of Sciences, 28-9,
  95, 141, 147, 148, 150;
  Revolution, 140-5, 149;
  telegraph systems, 91-2
Franklin, Benjamin, 231
Free radicals, 221, 225
Fremlin, J. H., 76
Frontinus, Julius, 179
Fuller, Rev. William, 164-6
Fulton, Robert, 232
Fungus, 237-8
Furneaux, Robin, 26, 28
Futurists, 189-90

Gagarin, Yuri, 236
Gagnan, Emile, 35, 36
Galápagos Islands, 122-6
Galaxy, 200; communication
  with aliens, 86-90; evidence
  of life in, 82-6
Galileo, 200, 230
Galileo, 200
Galois, Evariste, 145-51
Gama, Vasco da, 230
Gamow, Barbara, 80
Gamow, George, 80
Gases, 40-4, 83-4, 200, 201
Gaspra asteroid, 237
Genes, 221, 224, 225
Geometry, 146-8, 150-1
Germany, 157-8, 170, 171, 183,
  184, 185, 187
Gibbon, Edward, 12-13, 179
Giese, Tiedemann, 65
Giotto, 236
Glanvill, Dr Joseph, 156
Gold, 73
Gonorrhea, 154, 202
Good, I. J., 96
Goodwin, H. W., 233
Goodyear, Charles, 232
Gould, Stephen Jay, 128
Gramophone, 233; record, 235

Gravity, 77, 231, 234
Gray, Tom, 129-32
Greeks, 12, 178
Greenhouse effect, 85
Group theory, 145
Gunpowder, 229
Gutenberg, Johannes, 230
Hadar, 129-32
Hahn, Otto, 183
Halse, Fred, 49-56
Haldane, J. B. S., 201
Halley, Edmund, 72, 231
Halley's comet, 236
Halstead, Beverly, 135-8
Harding, Margaret, 237
Harrison, John, 173
Harvey, William, 231
Hawking, Stephen, 71
Heart, artificial, 236
Heat, 73, 141
Hebdomadal rule, 157
Helicopter, 235
Henry the Navigator, Prince,
   16-18, 19, 20, 229
Heraclitus, 66
Herschel, William, 231, 232
Herschel, Sir William, 108-10
Hertzsprung-Russell diagram,
   234
Hesse, 214-5
Heyerdahl, Thor, 235
Hipparchus, 66, 229
Hitler, Adolf, 157-8, 162, 183,
   187
Hominids, 131-4
Homo sapiens, 133
Hooke, Robert, 91, 231
Hourglass, 19
Howard, Dr Peter, 40-4
Hoyle, Fred, 80-1
Hubbard, Ron, 161
Hubble, Edwin, 234, 235
Hubble Space Telescope, 237
Hunter, John, 153-5, 202
Huxley, Thomas, 127-8
Hven, 70-1
Hydrofluorocarbons, 237
Hydrogen, 59, 181, 183, 219,
   231
Hypatia, 139-40
Hyperamnesia, 160
Hypnosis, 160

Icarus, 38-40
Ice Age, 30-33
ICI, 237
Iguana, land, 123-4
Immunization, 202
Incas, 22, 23, 28
Indigirka River, 30-2
Ingersoll, Simon, 233
Inquisition, 229
Intelligence, and
   communication systems,
   86-90

Iodine, 219, 220
Iodine-131, 60
Iridium, 136-7
Iron, 219
Italy, 170-2
Ivanovic, Vane, 211

Jacob, Henry, 167
Jacquard loom, 119
Jansky, Karl, 235
Jastrow, Robert, 136-7
Jefferson, Geoffey, 116-7
Jenner, Edward, 202, 232
Jet engine, 235
Johanson, Donald, C., 128-34
John of Austria, 99
Joint European Torus (JET), 237
Joliot-Curie, Frédéric, 184
Jupiter, 69, 83, 230, 231
Jussieu, de, 29-30

Kaffa tree, 168-9
Kahn, David, 99
Kahn, Herman, 189, 236
Kaldi, 167-9
Kelvin, Lord, 73, 233
Kennedy, President, 61, 236
Kepler, Johannes, 70, 230
Kettering, Charles, 234
Koert, Ulrich, 237
Kola Peninsula, 237
Kranz, Gene, 51

Lactantius, 67
Lagrange, Joseph, 145
Lalande, Joseph, 232
Lamont, Lancing, 186
Land of Cinnamon, 24
Laplace, Pierre Simon, 199-200
Large Electron Positron
   Collider, 237
Lavoisier, Antoine, 140-5, 231
Le Blon, Jakob, 231
Legasov, Valery, 60
Legendre, 146
Lehn, Jean-Marie, 237
Leibniz, Gottfried, 13, 72, 231
Leo X, Pope, 67
Ley, Willy, 33-5
Life: communication systems,
   86-90; evolution of, 83,
   122-8, in Galaxy, 82, 85;
   purpose of, 89
Light, speed of, 44, 74-80,
   173-4, 181, 231
Light-year, 75
Lightning conductor, 231
Linnaeus, Carolus, 203
Liouville, Joseph, 150
Lippershey, Johann, 230
Logarithms, 231
Lorentz, Hendrik, 78
Lousma, Jack, 50, 51, 53, 54-5

Lovelace, Ada, 118
Lovell, Captain James, 49-56
'Lucy', 128-34
Lull, Ram, 20
Luther, Martin, 67, 230
Lyell, Sir Charles, 214
Lysenko, Trofim, 234

Macaulay, Lord, 198
Mach, Ernst, 44
Machines, 72-3, 112, 176-9,
   229, 235
Madeira, 20
Magellan, Ferdinand, 20-22,
   230
Magnesium, 219
Magnetic field, 84
Magnetic tape, 235
Mammoth, 30-3
Man, evolution of, 223-4
Manganese, 219, 220
Maps, 19, 21, 230
Marat, Jean-Paul, 140-3, 144
Marcellus, 176-8
Mark, Dr Hans, 56
Mars, 231, 236
Marsh, Othniel, 135
Mars Observer, 238
Martin, Anthony, 83
Martin the Bohemian — *see*
   Behaim, Peter
Mary, Queen of Scots, 98,
   99-101
Mathematics, 66-7, 72, 146-50,
   169-72, 229
Maxwell, James Clark, 233
Maxwell's equations, 73
Mechanics, 175-8
Medawar, Peter, 135, 156
Medicine, 199, 202;
   experiments, 151-5
Mediterranean sea, 19
Megafauna, extinction of, 32
Meitner, Lise, 183
Mendel, Gregor, 232
Mercator, Gerardus, 230
Merlee, William, 229
Methane, 201
Mexico, 24
Mezzotint engraving, 198
Michelmore, Peter, 181, 185
Michelson, Albert, 74, 80, 233
Micro-time, 173
Microbes, 203-6
Milky Way — *see* Galaxy
Miller, William, 161
Mines, safety in, 201, 232
Miracles, 163-6
Missiles, ballistic, 188
Mistakes, scientific, 226-7
Molecules, 221, 223, 225
Molybdenum, 219, 220
Moncel, Du, 95
Monge, Gaspard, 200
Montgolfier brothers, 232

Moon: *Apollo 13* mission, 49–56; first landing on, 46–9, 235; geology of, 48; gravity of, 47; motion of 66, 70; radiation from, 217; relationship with Earth, 83–6
Moore, Patrick, 68
Moorhead, Alan, 122
Morley, Edward, 74, 80, 233
Morse, Samuel, 232; code, 102, 104
Motor car, 233
Motor cycle, 233
Munchausen, Baron von, 214, 216
Murphy, Bridey, 159, 160
Murphy's Law, 226–7
Murray, Lord George, 92, 94

Nairone, Antonius, 167
Napo River, 24–5, 26
Napoleon, 199–200
NASA, 44, 47–9, 53, 56, 57–8, 61, 136, 238
Natural selection, 125–8
Nature, La, 42
Nautilus, 236
Navigation, 171, 229, 230; discovery of Strait of Magellan, 21–2; early charts and instruments, 19–20; first attempts to discover America, 16–8; invention of marine chronometer, 173
Neanderthal Man, 133–4, 233
Nebular hypothesis, 200
Neptune, 232, 237
Neumann, John von, 117–9, 235
Neutron, 235
Newton, Isaac, 13, 28, 29, 71–2, 231
Nicholas of Cusa, 64
Nickell, Joe, 95
Nitrogen, 219, 231
Nixon, President, 46, 47, 48, 236
Nobel, Alfred, 232
Nollet, Abbé, 93
Northern Star, 19
Nostradamus, 230
Nuclear energy, 61–3, 181, 183, 184–5, 235, 237
Nuclear reactors, 58–9, 61, 237
Nuclear weapons tests, 225–6
Nucleohelicates, 237
Numbers, 170, 229

Oates, Captain Lawrence, 33, 34
Oppenheimer, J. Robert, 186, 235
Optical fibre cables, 237
Orellana, Francisco de, 26–7
Orestes, 139–40
Oroya fever, 151–3
Ovid, 38

Oxygen, 40–4, 83, 140, 141, 142, 219, 231
Ozone layer, 83, 237

Pacific Ocean, 21, 22
Palaeontology, 135–8
Paracelsus, 230
Parapsychology, 163, 164
Paris, wall of, 141, 142
Particle accelerator, 237
Particles, 173, 174, 183, 221, 223, 224, 225
Pascal, Blaise, 231
Pasteur, Louis, 193, 203–6, 233
Paul III, Pope, 64–7
Pauling, Linus, 225–6
Pearson, Jerome, 82–6
Penicillin, 235
Perpetual motion, 72–3
Peru, 23, 28, 151–3
Peugeot, 237
Phlogiston, 141–2
Phonograph, 95
Phosphorus, 219
Photography, 232
Photosynthesis, 232
Pi, 175
Pigafetta, Antonio, 21–2
Piltdown Man, 135
Pizarro, Gonzalo, 23–6, 230
Planets, 230, 231; conjunction of, 68; double, 83, 84–5; movement of, 64–7, 70
Plate tectonics, 84
Plato, 178, 228
Plummer, Mark, 164
Plutarch, 176–9
Pluto, 235
Poisson, Simeon, 148
Politian, 16
Pollio, Vitruvius, 175
Polo, Marco, 20
Portugal, 16–8, 20, 21, 229
Potassium, 219, 220–2, 223–5
Power, Henry, 156
Preece, Sir William, 95
Prescott, William, 23–6
Prevost, Pierre, 232
Priestley, Joseph, 231
Principia Mathematica 71–2
Printing, 231, 232
Probability, theory of, 231
Psychology, 160
Ptolemy, 16, 229
Pulsars, 237
Pythagoreans, 66

Quasars, 236

Radar, 158, 235
Radiation, 84, 217, 221–2, 225
Radio waves, 233, 235
Radioactivity, 59–60, 186, 218, 220, 221–2, 225, 233

Radium, 182
Railways, 232
Raleigh, Sir Walter, 27
Randi, James, 163–6
Rao, U. R., 84
Raspe, Rudolph, 214–7
Regis, Ed, 117
Reinhold, Erasmus, 68
Relativity, theory of, 74–80, 181–4, 234
Religion, 126–8
Richard, Louis-Paul-Emile, 146–7
Rockets, 44–5, 234, 235
Rogers Commission, 57, 58, 63
Romans, 11–3, 139, 176–9, 228
Romer, Olaus, 231
Ronalds, Sir Francis, 93–5
Roosevelt, President, 184–6
ROSAT, 237
Roscoe, Henry, 201
Royal Observatory, 72
Royal Society, 71, 72, 157, 198, 214–5, 231
Rubber, 28, 232
Rupert, Prince, 198
Russell, Bertrand, 74
Russia, 162, 188
Rutherford, Lord, 183, 231
Ryle, Martin, 80–1

Sachs, Alexander, 185–6
Sachs, Julius, 232
Sagan, Carl, 82, 236
Sambuca, 177
Samuel, A. L., 119
Sanyo Electric, 238
Satellites, 235, 236, 237
Saturn, 69, 70, 83
Schonberg, Nicholas, 65
Science: beginnings of, 12–3; classical theory of, 178, 228; and credulity, 162–3; divisions in, 135–8; examinations, 119–20; explosion in seventeenth century, 198–9; incompetence in 96–8, 226–7; influence of Aristotle, 156–7; and religion, 127–8, 180; role of imagination, 190–3; school textbooks, 206–11
Science fiction, 189–97
Scorpions, 177
Scott, Robert Falcon, 33–5
Second Law of Thermodynamics, 73, 233
Second World War, 99
Semiconductors, 238
Senièrgues, Dr, 29
SETI programme, 82, 86, 87
Shakespeare, William, 21
Sharks, 211–3

Siberia, 30-3
Siebe, Augustus, 35
Sivel, Theodore, 40-2
Slattery, Bart, 45-6
Slizard, Leo, 184, 185
Sloane, Hans, 199
Smallpox, 202, 232
Smoot, George, 237
Socrates, 139, 178, 228
Sodium, 219
Solar System, 84
Solar Year, 66
Solomon's House, 192-7
Sophocles, 56, 63
Sound, speed of, 44-5
South America, 20-30, 151-3
South Pole, 33-5
Soviet Union, 56, 58-60, 62, 63,
  234, 236
Space: *Challenger* explosion,
  56-8, 60-3; early exploration
  of, 40-4; Earth's movement
  through, 74, 77; first man in,
  236; first Moon landing, 46-9;
  Apollo 13 Moon mission,
  49-56; *Galileo* probe, 237;
  Giotto probe, 236; *Mars
  Observer* probe, 238;
  shuttles, 236; *Voyager 2*
  probe, 237
Space travel, 75-80, 82
Spain, 21, 22-7, 93
Spallanzani, Lazzaro, 193
Species, variation of, 124-6
Spectacles, 229
Spontaneous generation, 193
Spruce, Richard, 27
Stars, 19, 66, 69, 70, 73, 201,
  229, 232, 234
`Steady state' universe, 80-1
Steam engine, 231
Stefan-Boltzman law, 217
Stethoscope, 232
Stjerneborg Observatory, 70-1
Strait of Magellan, 21-2

Submarine, 232
Sulphur, 219
Sun, 64-7, 70, 181, 217
Superconductivity, 234
Superconductors, 238
Supernova, 69, 201
Surgery, transplant, 154
Swetz, Frank J., 170
Swigert, Jack, 49-56
Syphilis, 154-5, 234
Syracuse, siege of, 176-8

Tachyon, 77
Tax farming, 141, 143-4
Taylor, Stuart Ross, 80
Technology, 56, 60, 62, 94
Telegraph, electric, 91, 92-4,
  232
Telephone, 94-5, 96, 233
Telescope, 230, 235
Television, 234
Terquem, 146
Three Mile Island nuclear
  accident, 60, 61
Tichborne, Roger, 116
Tides, 83, 84
Time: units of, 173
Time travel, 75-80
Tissandier, Gaston, 40-4
Tortoise, giant, 123
Trace elements, 218-25
Transistors, 190, 235, 238
Trevithick, Richard, 232
Trigonometry, 230
Turing, Alan, 112-9, 235
Twin Paradox, 75-8
Tycho's Star, 69

UFOs 160, 161
Uranibourg Observatory, 70-1
Uranium, 59, 184-5
Uranus, 231
Ussher, Bishop James, 230

Vaccines, 202, 232
Velikovsky, Immanuel, 161
Ven — *see* Hven
Venereal disease, 154-5, 202
Venice, 171
Venus, 231, 236
Vernam, Gilbert, 101, 102, 108
Verruga peruana, 151-3
Versi, Piero de, 171
Vigenère, Blaise de, 99
Vinci, Leonardo da, 230
Virgo, galaxies, 89
Viruses, 233
Voyager 2, 237

Wallace, Alfred Russel, 32
War, 176-8, 228
Water, 142
Watson, James, 235-6
Watson, Thomas, 94-6
Watt, James, 231
Weather forecasts, 229
Weintraub, Joe, 119
Wells, H. G., 47, 76
White, Edward, 236
Whitman, Walt, 73
Wilberforce, Bishop, 127-8
Wiles, Andrew, 238
Wilson, Dr E. A., 33
Wolff, Heinz, 119
Wright brothers, 234

X-rays, 233

Yeager, Charles 'Chuck' E.,
  44-5
Yeasts, 204-5

Zenith, 40-4
Zinc, 219, 220
Zirconium, 59
Zweig, Stefan, 13

.